THE BIOTECH PRIMER ONE

The Science Driving Biopharma Explained

An Insider's Guide to the Science Driving
the Biopharma Industry for the Non-Scientist

THE BIOTECH PRIMER ONE
The Science Driving Biopharma Explained

Published by:
Biotech Primer Inc.
8600 LaSalle Road, Suite 605
Towson, MD 21286

BiotechPrimer.com
info@BiotechPrimer.com
Copyright @ 2019 Biotech Primer Inc., Towson, Maryland, USA
Published by Biotech Primer Inc., Towson, Maryland, USA

All Rights Reserved. No part of this publication may be reproduced, stored in a retrieval system or transmitted in any form or by any means, electronic, mechanical, photocopying, recording, scanning or otherwise. Requests to the Publisher for permission should be addressed to Biotech Primer Inc., 8600 LaSalle Road, Suite 605, Towson, MD 21286 USA, or emailed to info@BiotechPrimerInc.com.

Trademarks: Biotech Primer Inc. logo, Life Science Training From Industry Experts, Biotech Primer and related marks of Biotech Primer Inc. registered in the United States may not be used without written permission. All other trademarks are the property of their respective owners. Biotech Primer Inc. is not associated with any product or vendor mentioned in this book.

Biotech Primer Inc. makes no representation of warranties with respect to the accuracy or completeness of the contents of this work. Biotech Primer Inc. shall have exclusive ownership of all copyrights, trademarks and other intellectual property in connection with the hard copy handout and all other materials in any form whatsoever given to reader.

This work is sold with the understanding that the publisher is not engaged in rendering legal, scientific or other professional services. The publisher shall not be liable for damages arising herefrom. The fact that an organization is referred to in this work as a citation and/or a potential source of further information does not mean that the publisher endorses the information the organization may provide or recommendations it may make.

Biotech Primer Inc. also publishes its books in a variety of electronic formats. Content that appears in print is also available in electronic books.

ISBN 979-8-7042-3240-7 (softcover)
ISBN 978-1-5136-5505-5 (eBook)

THE BIOTECH PRIMER ONE

Biotech Primer Inc.

Dear Reader,

Welcome to the second edition of Biotech Primer One: The Science Driving Biopharma Explained. Like the first edition, this second edition is intended to be a foundational science guide for non-scientists who work in the biopharma industry. Biotech Primer One has updated examples, additional material, and has been reedited for greater clarity. I hope this book will serve as a jumping-off point for your biopharma career. I encourage you to continue your biotech education by reading the sequel, Biotech Primer Two: Targeted Biologics Explained, which outlines how novel therapies cure disease with the help of our human immune system. Together, both books provide the background required to communicate more effectively with scientists, clients, and colleagues.

In addition to our two books, I encourage you to continue your biotech education by participating in any of our online courses and by reading our WEEKLY e-newsletters. The WEEKLY explains the science behind the biotech news headlines and hundreds of WEEKLYs can be accessed under Publications at biotechprimer.com.

Happy Learning!

Stacey Hawkins

Stacey Hawkins
Founder, Biotech Primer Inc.

ABOUT BIOTECH PRIMER

Biotech Primer Inc. delivers customized training to help professionals understand the science, business, and regulatory processes essential to the biopharmaceutical, medical device, and molecular diagnostic sectors. Our industry experts develop and deliver customized courses, so you are assured our content is always up-to-date and relevant to your training needs. .

When we opened Biotech Primer's doors in 2001, our sole focus was entry-level science education for the non-scientist, which helped clients bridge in-house communication gaps and develop more cohesive, well-trained teams for product development. Since then, we expanded to offer advanced training for all levels of experienced professionals, including scientists—an estimated 100,000 people trained worldwide to date. We help prepare companies to make strategic business decisions, navigate important regulatory hurdles, and move healthcare products from the bench to the bedside. To accomplish these goals, we offer a diverse range of learning, ensuring participants retain and put into practice what they learn.

WE OFFER:

- **Customized Training:** Organizations can choose from our master courses—featuring core material from our extensive library of life science topics—or let us fully customize a course with completely original content specific to your organizations needs.

- **Live, Online Courses:** Register to take our prescheduled, one-day or two-day master classes offered throughout the year via Zoom.

- **On-Demand, Online Learning:** Learn anywhere, anytime , at your own pace from our library of 60-minute on-demand classes. Designed for individuals, bulk purchase discounts available for organizations.

- **The Biotech Primer WEEKLY e-Newsletter:** Hundreds of one-pagers explaining the science behind today's headlines available. Head over to Publications on our website and take advantage of our original content.

biotechprimer.com

PUBLISHER'S ACKNOWLEDGEMENTS

We are proud of this book and welcome your feedback at info@BiotechPrimer.com.

For more information on the basic science that drives the biopharma industry read our WEEKLY e-newsletter at WEEKLY.BiotechPrimer.com and follow us on twitter @BiotechPrimer.

For a list of live and on-demand online courses visit us at BiotechPrimer.com

Author: Emily Burke,
Ph.D. Editor: Sarah Van Tiem

Illustrations: Michelle Leveille
Cover & Book Layout: Tara Price

USER'S GUIDE

Throughout the book, you will see a variety of callouts, each designed with a specific purpose in mind. Here is a guide to the intent of each:

INDUSTRY NOTE: Explains how the biopharma industry uses science to advance healthcare.

BIOPHARMA INNOVATION: Examines a specific product, explains how it works and relates it to the science described in the main text.

COCKTAIL FODDER: A Biotech Primer classic, pull them out the next time you are stumped for conversation at your company holiday party!

GOING FURTHER: Takes a more in-depth look at a topic related to the main text.

TRICKY TERMS: Highlights the meaning and usage of industry terms that are especially important, or that tend to be confusing to non-scientists.

GLOSSARY: All **bolded and underlined** words in the text are found in the glossary at the back of the book.

Happy reading!

CONTENTS

CHAPTER 1: PUTTING THE BIO IN BIOTECH — 1

Cells: The Basic Unit of Life	1	Growth Factor Signaling	12
Cell Structures	5	G Protein-Coupled Receptor Signaling	15
But What Do Cells DO?	7	Ion Channel Signaling	16
Talk Talk	7	A Synchronized Tangle of Communication, Not a Line	19
Grow Cell Grow	9		
Cell Signaling	11		

CHAPTER 2: BLUEPRINT OF THE CENTURY: DNA — 21

Decoding the Master Plan	21	Putting It All Together	26
The Story of DNA	22	How DNA Replicates	29
Life's Building Blocks	24	Organization of DNA	32

CHAPTER 3: GENE EXPRESSION: FROM GENE TO PROTEIN   35

Step One: Transcription	35	Protein Therapeutics	41
Codon Table	37	Peptide Therapeutics	43
Step Two: Not Lost in Translation	38	Protein Shape and Sickle Cell Disease	44
Proteins Take Shape	39		

CHAPTER 4: GENETIC VARIATION   47

Getting Snippy	51	Cancer: A Polygenic Disease	57
Genetic Basis of Disease: Monogenic Disease	52	Pharmacogenomics	58
		Benefits of Pharmacogenomics	59
Genetic Basis of Disease: Polygenic Disease	54	Precision Medicine in Practice	59

CHAPTER 5: DECIPHERING GENETIC VARIATION   65

Our Secret Code	66	A Chip Off the New Block: SNP Chips	76
Cracking the Code: The Human Genome Project	66	Chips at Work: Direct-to-Consumer Genetic Tests	79
Pieces of You	68		
Sequencing by Synthesis: "Hi-Seq"	70	Clinical Studies	80
Machines That Read: Third Generation	71	A Dash of Difference	82
		Alphabet Soup: SNPs and STRSs	82
Application: Deciphering a Killer	74	This DNA is Not That DNA	83
Easily Confused: Genome Sequencing vs. SNP Genotyping	75	Deep Ancestry	84

CHAPTER 6: GENETIC ENGINEERING — 87

Transgenic Organisms	88	Engineering Mammalian Cells	96
Transgenic Organisms: What Don't They Do?	88	Recombinant Proteins in Healthcare	99
		Animal Farm	100
Bacterial Scissors?	90	Getting the Goat	100
Bacteria Factories	93	Magic Mice	102

CHAPTER 7: DRUG DISCOVERY — 103

Identifying the Target	107	Glowing Results	114
Types of Targets	108	High Throughput Screening	117
Validating the Target	109	Combinatorial Library Success	118
Therapeutic Choices: Small or Large?	111	From Hit to Lead	120
		Biomarkers	122
Assay Development	113		

CHAPTER 8: DRUG DEVELOPMENT: FROM THE LAB TO THE CLINIC — 125

Regulatory Agencies	125	More Paperwork: Patents	134
Regulatory Step 1: Preclinical Trials	126	That Special Something	135
Regulatory Step 2: Clinical Trials	128	What's in a Name: Generic	137
New Kid In Town: Adaptive Design	132	What's in a Name: Biosimilar	137

CHAPTER 9: BIOMANUFACTURING — 139

The Big Biomanufacturing Picture	140
Bigger Better Faster More!	142
Full Speed Ahead	144
Campaign Season	144
Cell Banks	145
Scale-Up and Manufacturing Process	146
Harvest Time	148
Column Chromatography	148
More Chromotography: Ion Exchange	150
Still More: Affinity Chromatography	151
And One More: Size Exclusion Chromatography	151
Formulation, Fill…	152
…and Finish	152

Conclusion — 153

Glossary — 155

CHAPTER 1

Putting the BIO in Biotech

At its most basic level, biology seeks to understand how cells function. What do they do and how do they do it? Biotechnology applies this understanding toward developing new products—from cancer treatments to biofuels. Later chapters will examine some of these literally life-changing products. But first, we'll build our understanding of life at its most fundamental.

Cells: The Basic Unit of Life

All living things are made of cells. Organisms can be made of a single cell or in the case of human beings—of *trillions* of cells. **Unicellular** organisms include yeast, bacteria, and some algae. **Multicellular** organisms are land plants, animals, and people. We're intimately familiar with multicellular organisms—our families or coworkers, our pets and the trees, grass and flowers around us. However, it's the unicellular organisms that are the tiny, essential motors which in large part drive the biotechnology industry.

Chapter 1: Putting the BIO in Biotech

Cocktail Fodder

Ostrich eggs are the world's largest single cells, weighing up to 3 pounds.

Organisms consist of two distinct classes of cells: prokaryotic (**prokaryotes**) and eukaryotic (**eukaryotes**). You likely haven't heard of either—but they're everywhere! First, the prokaryotes. These are bacteria. Most everything around us is covered in them—doorknobs, cellphones and don't even think of kitchen or bathrooms. No need to fret though—most bacteria doesn't cause disease. In fact, many are beneficial. Just think of the little dudes that live inside us and aid with digestion. The collection of microbes that live in and on our body is referred to as the **microbiome**. In recent years, scientists have discovered that the microbiome plays a critical role in human health, and not just in digestion. Our microbiome also helps regulate immunity, inflammation, obesity, and even neurological health. Want to know more about our amazing microbiome? Look for *The Biopharmaceutical Primer: An Insider's Guide to Today's Advanced Therapies,* coming soon. Our new publication takes a more in-depth look at how the human body naturally fights disease and how the biopharma companies are modulating it to develop new treatments.

The eukaryotes are everything else—all non-bacterial cells.

So where do the funky names come from? Even though this book covers biotech, here we take a brief interlude into language. And, as is so often the case in science, blame the Greeks. *Karyote* comes from the ancient Greek word for nut or kernel. For later scientists, this term designated the cellular nucleus. The prefix *pro* simply means "before." Thus, prokaryote means a cell before, that is without a nucleus.

The prefix *eu* means "true." All eukaryotic cells contain a true nucleus. Every cell in your body—and in all other multicellular organisms too—is eukaryotic.

Cocktail Fodder

Red blood cells are the only human cells that do not have a nucleus.

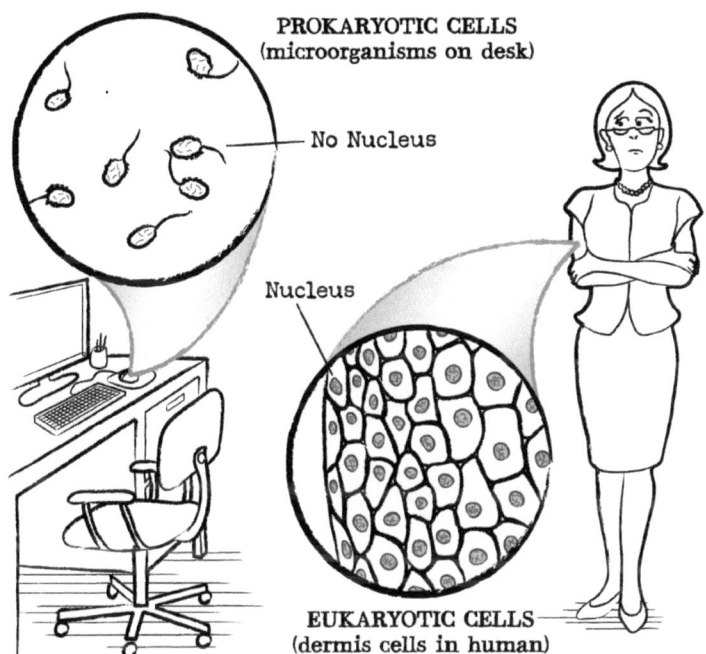

Cocktail Fodder

Human cells have various lifespans. Skin cells last 14 days, red blood cells last 120 days, and neurons last for an individual's entire lifespan.

GOING FURTHER

An adult human is made up of roughly 100 trillion cells. That's 100 million million of more than two hundred different kinds. The multifarious human cell types include organ-specific, muscle, and nerve that all developed from a single miniscule fertilized egg. Each type of cell carries out a specialized function. Somehow, they all work together 24 hours a day, seven days a week, 365 days a year throughout our lives. Go cells!

Chapter 1: Putting the BIO in Biotech

INDUSTRY NOTE: E. coli

A common bacterium found in our intestines, *Escherichia coli* (*E. coli*), is the microbial workhorse of research labs. Because *E. coli* has been so thoroughly studied and is FDA-approved for the production of human therapeutics, it's the most widely used prokaryote in the biotech industry. You may have heard that *E. coli* causes diarrhea, urinary tract infections and other unpleasantness. However, only a few bad actors cause intestinal mayhem. Most strains are perfectly harmless.

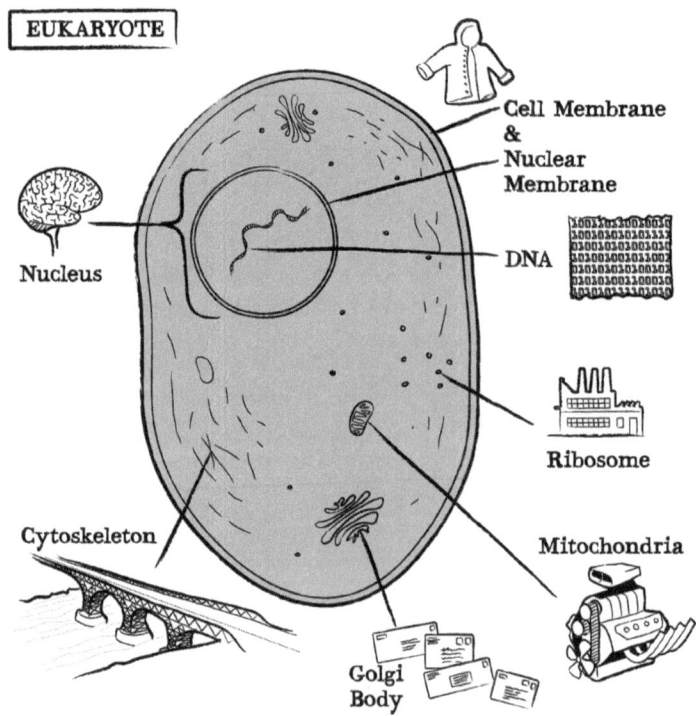

Cell Structures

Now that we've laid out some cell basics, let's take a peek under the hood. We'll look at the structures, known as **organelles**, found within cells and their functions.

All cells are surrounded by a **cell membrane**. This keeps what's inside the cell *inside* and the stuff that surrounds the cell *outside*. The membrane also acts as a doorkeeper, allowing in certain molecules. For instance, sugars, which provide energy, are allowed to enter. Equally important, the membrane blocks other molecules, such as those with a positive or negative charge. This enables the cell to maintain the specific chemical environment required for optimal functioning. The doorkeeper also allows waste products out. The membrane is also very flexible, allowing cells to change shape and move freely. Some cells, such as bacteria and plant cells, also have **cell walls**, which make them even stronger.

In prokaryotes, genetic material, or **DNA**, floats freely around inside. In the eukaryotes—yeast, invertebrates, us—however, DNA is nestled more safely in the cellular nucleus. DNA provides the blueprint for making the proteins that perform all cellular functions. The cell nucleus has its own protective barrier, the **nuclear membrane**, which further safeguards the enclosed DNA.

Proteins make most everything happen inside cells. And, as it so happens, many biotech products *are* proteins. These microscopic worker bees are made by **ribosomes**.

Cocktail Fodder

Theoretically, the human body may be incapable of living beyond 120 years due to genetic coding which is thought to limit the amount of times our cells can divide. The longest documented human lifespan is that of Jeanne Calment of France (1875–1997), who died at age 122 years, 164 days.

Cocktail Fodder

Mitochondria have their very own DNA. 100% of our mitochondrial DNA comes from our mothers. Thanks, Mom.

Making proteins, allowing molecules to cross the membrane—any cellular business—requires energy. The fuel comes from **mitochondria**, which take the energy from sugars and fats and convert it into a useable form—a molecule called **ATP**.

Cells manufacture thousands of different proteins. Many of them need to get to a particular place in the cell to do their job. A structure called a **Golgi body** acts as a post office—sorting each protein and sending it to the correct destination.

There's a lot going on in one little cell. Organelles—the tiny membrane-bound cellular compartments such as mitochondria—and proteins need to move within the cell. With so much traffic, there must be a freeway, right? That freeway is an extensive network of tubes and filaments known collectively as the **cytoskeleton**. This structure coordinates and carries out the cell's transportation needs. The cytoskeleton does another vital job as well—it provides a scaffolding for the cell membrane—literally, a cellular skeleton.

INDUSTRY NOTE: Mitochondria

It makes sense that heart and other muscle cells contain lots of mitochondria since they need lots of energy to function well. Our brains also require boatloads of energy. A whopping 20% of the sugar we take in feeds our noggins. Researchers have found the brain cells of people with neurodegenerative diseases such as Parkinson's and Alzheimer's contain fewer mitochondria than those of healthy people. This observation opens up the possibility of a new therapeutic approach—developing drugs that preserve brain cell mitochondria.

Chapter 1: Putting the BIO in Biotech

But What Do Cells DO?

The precise answer to that question varies with cell type. Heart cells pump blood, liver cells detoxify chemicals, pancreatic cells make insulin, to name a few. However, all cells, prokaryotic and eukaryotic alike, carry out four basic functions: communication, growth, division, and protein manufacture.

Cocktail Fodder

The longest cells in the human body are the motor neurons. They can be up to 4.5 feet (1.37 meters) long and run from the lower spinal cord to the big toe.

Talk Talk

In multicellular organisms, cells communicate. Since cells don't have mouths, ears, or email access, they rely on chemical messenges. For example, one cell releases a chemical message, say a hormone, to a second cell. The second cell is called the target. The target receives the message through proteins inserted into its membrane known as **receptors**. These protein receptors control the flow of information across the membrane. When the chemical message (also known as a signaling protein) binds to the target receptor, the receptor changes shape, and **transduces**—converts from one shape to another—which allows the chemical message to cross the cell membrane. This process of cellular communication is known as **signal transduction**. The most common result of signal

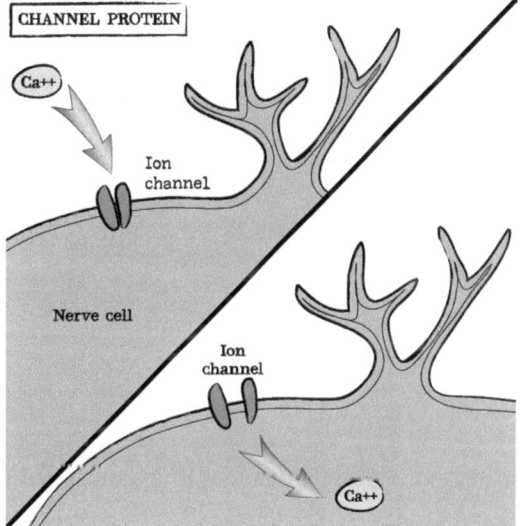

The Science Driving Biopharma Explained

transduction is switching protein production on or off. This is more often called **gene expression**. More on this key step in cell decision-making in later chapters.

Channel proteins are also involved in cellular communication. They're the gates that allow ions—positively or negatively charged atoms—such as sodium (Na^{++}) to cross the cellular membrane in response to stimulus. In neurons, ion transport between cells serves as a principle means of signal transduction. The influx of calcium ions (Ca^{++}) into a nerve cell releases **neurotransmitters**—chemical messengers specific to the nervous system. Different neurotransmitters regulate different brain functions, including muscular movement, memory, learning, and mood regulation. We'll examine communication between nerve cells in more detail shortly.

GOING FURTHER: Insulin

The regulation of blood sugar by the protein hormone insulin offers a basic example of cellular communication. After a meal, beta cells in your pancreas sense increased blood glucose. In response, they release insulin into the bloodstream. Insulin molecules attach to specific insulin receptors on muscle cells. This delivers a signal to the inside of the muscle cell to send proteins to the cell's surface that transport glucose from our blood into the cell, resulting in glucose uptake. As a result, blood glucose levels remain constant.

Grow Cell Grow

All cells can grow. When one does, it makes more organelles, such as mitochondria, to keep up with energy demand. The growing cell also produces proteins. In fact, they make up most of the pumped-up cell's increased mass.

Before a cell divides, it copies its entire DNA sequence, also known as its **genome**. One copy goes into each new "daughter cell." Each daughter cell possesses a copy of the original genome and about the same number of organelles as its parent. The daughter cell also eventually grows to the same size as the original.

Cell growth and division are tightly linked and strictly controlled. Some cells, like those in the skin or lining the gut, grow and divide often. Most cells in a mature organism, however, do so rarely. Cells grow and divide *only* if they receive a signal—a **growth factor**—from *outside*. When growth factors are released and attach to the growth factor receptor in the cell membrane, a cascade of signals "activate" the cell's DNA. This kicks off the production of proteins required for cell division.

Some cells continue to divide without a growth signal. When these unruly cells continue to flourish in an uncontrolled manner, they can become cancerous. See the section *Cell Signaling* on page 11 for more on the signaling pathways involved in cell growth and division.

Cocktail Fodder

Ribosomes can assemble an average-sized protein in about a minute.

Chapter 1: Putting the BIO in Biotech

Cocktail Fodder

Ribosomes can assemble an average-sized protein in about a minute.

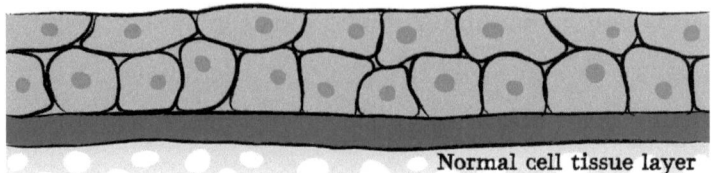

Normal cell tissue layer

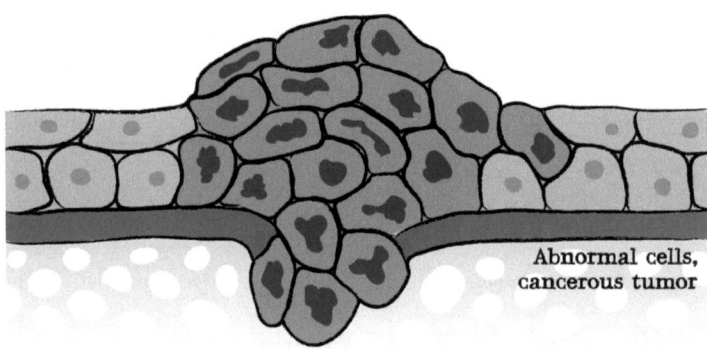

Abnormal cells, cancerous tumor

BIOPHARMA INNOVATION: Cells Are Nature's Makers

From a biotech perspective, one of the key properties of cells is the ability to manufacture biologically important molecules, especially proteins. As mentioned earlier, that's because many biotech products are actually cellular proteins made by ribosomes. The "cookbook" for proteins come from the cell's DNA. With sufficient energy and information, ribosomes get busy. Many proteins physically interact with others to form functional multi-protein complexes. They handle almost all critical cellular functions—from chemical reactions to communicating with the outside world.

Cell Signaling

Some cells send signals while others receive them. Most cells do both. There are two kinds of signals: **chemical hormones**, such as adrenaline, and **proteins**, such as insulin. They're produced within specialized cells (the signaling cell) and released to find their target cells. A signal is often called a **ligand**. Sometimes, the signaling and target cell are one in the same.

The target cell is sometimes in direct contact with the signaling cell. If so, the signal can travel by diffusion through the intracellular space. A target can also reside in a different part of the body and receive its signal through the bloodstream. No matter where, after receiving a signal, the target cell's response stems from the nature of the signal itself.

There are three key types of cell signaling: **growth factor signaling**, **G-protein coupled receptor signaling**, and **ligand-gated ion channel signaling**.

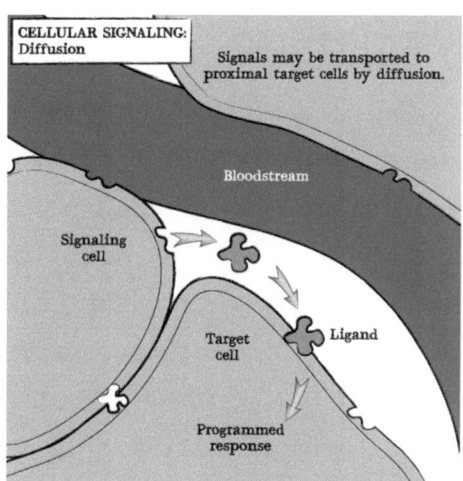

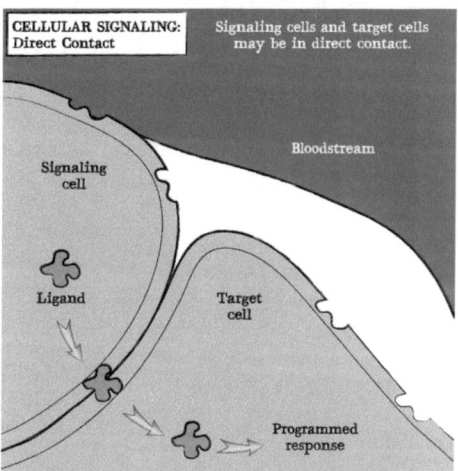

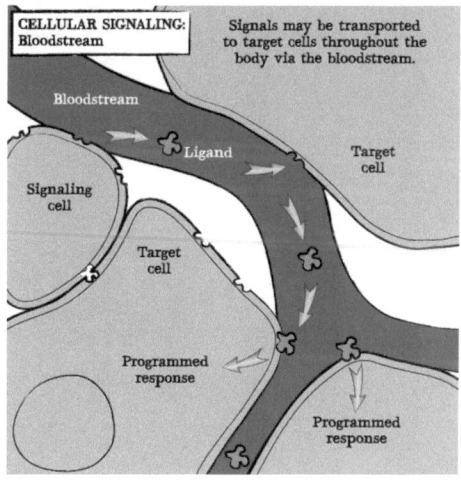

Growth Factor Signaling

Cells are constantly exposed to various growth factors. Growth factors are proteins that tell a cell to multiply. Growth factors that cells respond to depend on the cell surface receptors. For instance, one of the most well-researched factors, epidermal growth factor (EGF), stimulates the production of skin cells during wound repair. Skin cells, as well those covering the gut, lung and breast, have or express receptors for EGF. Other cells have receptors for different growth factors. For example, nerve cells express receptors for nerve growth factor (NGF).

After a cell receives the initial growth factor signal, the enzymatic activity of the internal portion of the growth factor receptor is activated. That switches on protein kinase activity. This is the ability to transfer a phosphate group from one molecule to another. Scientists sometimes refer to these receptors as receptor tyrosine kinases (RTKs), because the RTK moves a phosphate to the amino acid tyrosine found on the recipient protein.

This transfer, in turn, causes the protein receiving the phosphate group to change its shape slightly. That alteration typically turns on the recipient protein's own kinase activity. This newly-activated protein kinase then "turns on" yet another kinase protein and this happens multiple times down the line. Biochemists call this complicated dance of proteins a **signal transduction cascade**.

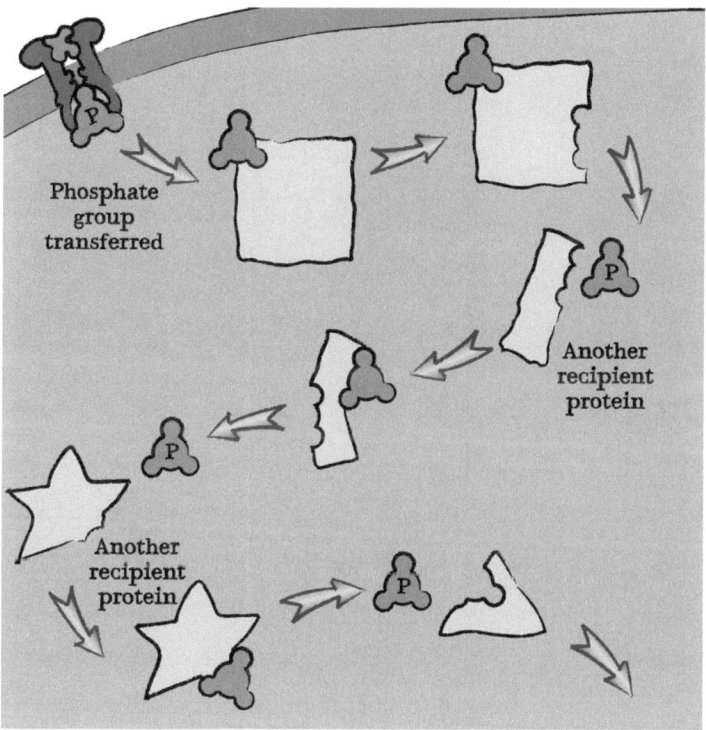

The last element to be phosphorylated is usually a transcription factor. Once that happens, this protein enters the nucleus. There it binds to the DNA at a particular spot ultimately turning on the expression of a specific gene.

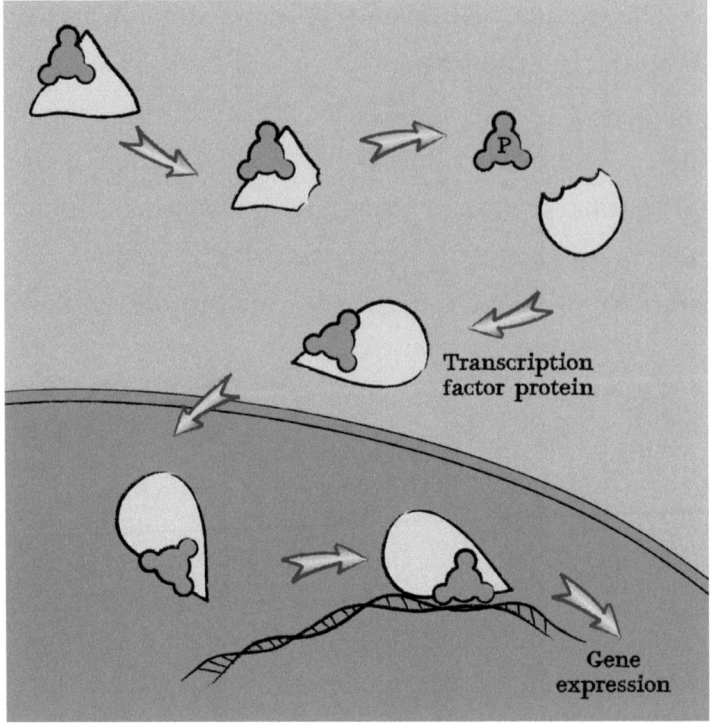

Defects in growth factor signaling are associated with various cancers. A major challenge in oncology lies in more fully understanding the complicated communication that initiates cell division. Further insight into these processes will fuel the development of targeted therapies for specific cancers. Defects in growth factor signaling are associated with various cancers. A major challenge in oncology lies in more fully understanding the complicated communication that initiates cell division. Further insight into these processes will fuel the development of targeted therapies for specific cancers.

G Protein-Coupled Receptor Signaling

G protein-coupled receptors (GPCRs) are another common receptor molecule. Almost all cells use them to regulate processes from blood pressure to nerve transmission to stomach acid secretion. Their name comes from the fact that the receptors are coupled to a signaling protein called a **G protein**. GPCRs are more complicated than growth factor receptors in that they travel back and forth across the cellular membrane seven times. That movement gives these receptors their alternate monikers: seven-transmembrane or serpentine receptors.

Upon ligand binding, the GPCR undergoes a subtle conformational (shape) change which causes it to interact with the associated G protein. This results in some G proteins leaving the complex and interacting with other cell proteins. Some of these other cellular proteins are kinase proteins, as described for growth factor receptor activation.

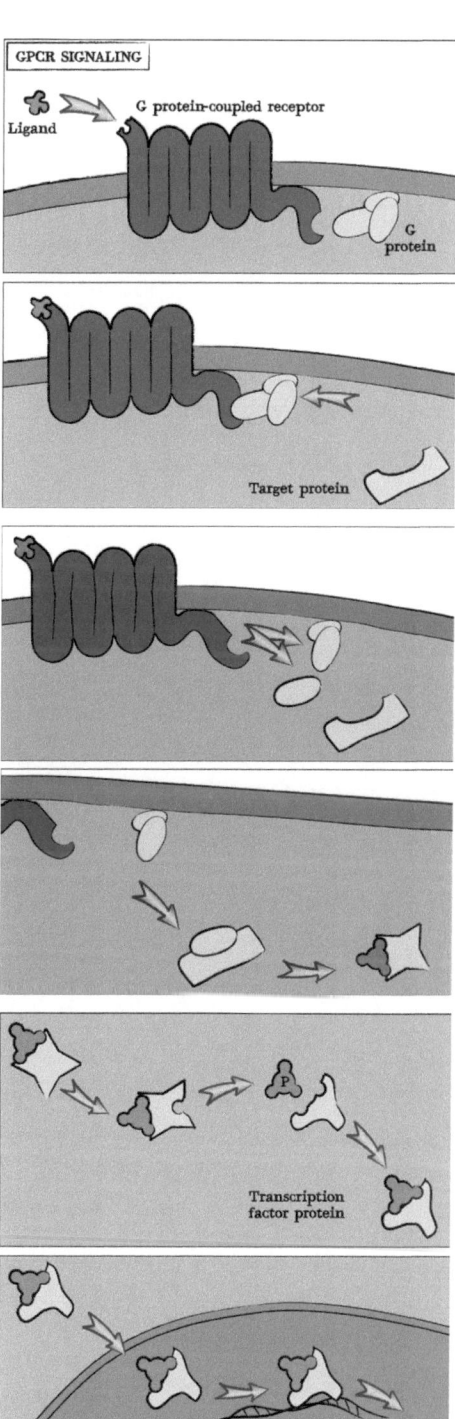

INDUSTRY NOTE: Drug Targets

A variety of ligands activate GPCRs—simple chemicals such as epinephrine and histamine or more complex signaling proteins. The results of GPCR activation vary, ranging from the activation of gene expression to the release of an already-made protein to the reorganization of the cytoskeletal proteins. This assortment of responses and the critical role these receptors play in countless biological processes make them popular drug targets. They're the focus of an estimated 25 to 40% of biotech drugs in development or on the market.

Ion Channel Signaling

Certain cells, commonly called **excitable cells**, are unique because of their ability to generate electrical signals. Although several types of excitable cells exist, including neurons and muscle cells, all of them use ion channel receptors. Ion channel receptors are located in the cell membrane, arranged so that it forms a passageway extending from one side of the cell membrane to the other. These passageways, or ion channels, open and close in response to chemical signals. When an ion channel is open, ions (positively charged or negatively charged atoms) move into or out of the cell.

Excitable cells maintain a different ion concentration level in its cytoplasm than exists on the outside in the extracellular environment. Together, these concentration differences create a small electrical potential across the cell membrane. When concentration differences are right, ion channels open and allow rapid ion movement into or out of the cell, and this movement creates an electrical signal.

Let's take a closer look at neuron signaling. When the concentration gradient is optimal, ions channels open and ions enter and travel along the neuron. The ions create an electrical signal that moves down the cell until it comes to the cell's end. When electrical signals reach the end, it triggers the release of chemical messengers called neurotransmitters because the electrical signal can't "jump" the neural synapse space.

The neurotransmitters release their content into the neural synapse—the space between nerve cells. Then they bind receptors in the next neuron, allowing ions to enter and start the process over.

It's important to switch off nerve impulses once they're transmitted. Consequently, the neurotransmitter gets rapidly removed from a synapse by specific reuptake channels in the presynaptic neuron. This shuts down the signal until another nerve impulse arrives. Blocking reuptake channels with the right molecules therefore sustains and amplifies neuronal signaling. This can significantly influence neural function.

Various molecules and receptors involved in this process are excellent drug targets. Reduced levels of the neurotransmitter serotonin are linked to depression. This observation formed the basis for a highly successful class of antidepressants called selective serotonin reuptake inhibitors (SSRIs). These drugs include Prozac, Zoloft and Paxil. They block reuptake of serotonin, which increases its concentration between synapses and often helps relieves depression.

Some recreational drugs also act as reuptake inhibitors. Cocaine, for example, blocks the reuptake of the "feel good" neurotransmitter dopamine.

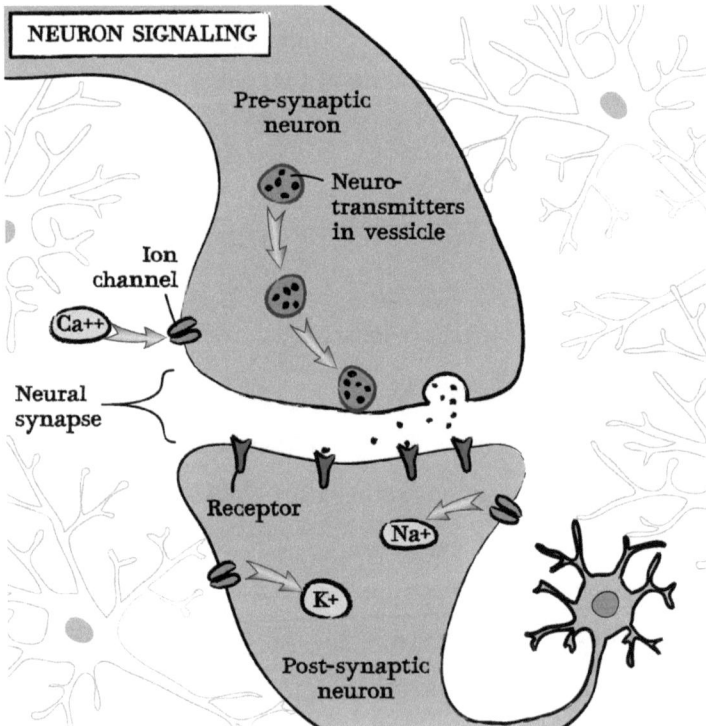

A Synchronized Tangle of Communication, Not a Line

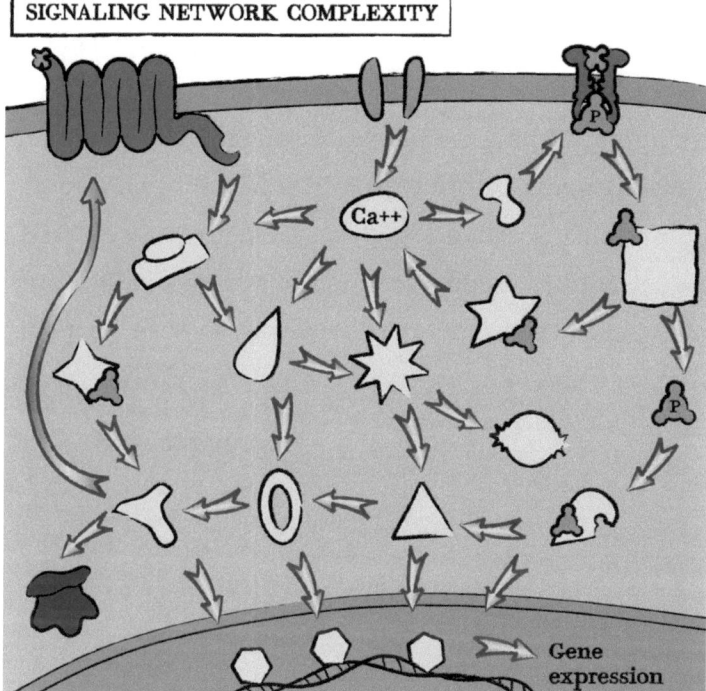

Cocktail Fodder

DNA is the largest molecule of the cell.

This book depicts signaling through linear, individual pathways. However, the reality is hugely complex. Hundreds of different receptor types exist on the surface of every cell. Each of these binds a different ligand (including growth factors, hormones, and neurotransmitters). They include receptor tyrosine kinases, GPCRs and ion channels. Some receptors, like GPCR, bind more than one ligand. Ligand binding to each receptor starts a different signaling cascade, ultimately regulating gene expression and cellular function.

Signaling pathways aren't isolated; different pathways can overlap. For example, signaling proteins from GPCR and RTK pathways often interact. The result is the activation of a variety of **transcription factors** leading to gene expression. This gene expression response is dependent on the strengths of each extracellular signal. Cellular communication gets even further complicated because signaling networks affect processes beyond gene expression. Signaling networks play a role in protein synthesis, the secretion of cellular protein, and activities in the mitochondria and cytoskeleton to name only some of what's up every moment of every day in our busy, busy cells.

In Chapter 2, we go even further into the stuff of life. Get ready for DNA!

CHAPTER 2

Blueprint of the Century: DNA

Decoding the Master Plan

At the dawn of the 21st century, the Human Genome Project had just been completed. This revolutionary undertaking—determining the exact size, sequence, and location of genes within the human genome—is one of biology's greatest achievements. For the first time, researchers possessed the complete human blueprint. Ideally, they could now pinpoint the source of countless diseases. This ushered in what many, including the European Commission and the National Academy of Sciences, have referred to as the Century of Biology. Since then, the ability to quickly and economically sequence the human genome has increased exponentially.

DNA is the blueprint of this Century of Biology. Found in all living cells, DNA is the genetic material that stores and transfers information. In this chapter, we look deep inside the cell and examine the molecule upon which the entire biotechnology industry rests, starting with key discoveries and the scientists behind them.

The Story of DNA

In the 1850s, an Austrian monk named Gregor Mendel performed breeding experiments with pea plants. He observed that certain genetic characteristics were passed down from one generation to the next in specific ratios. Because he was the first to analyze the inheritance of traits systematically, many consider Mendel the father of genetics. Mendel's contributions are even more impressive because though he didn't know about DNA, he predicted its existence. He called what we know now as DNA "particles of inheritance" which he suspected were responsible for passing traits from generation to generation.

Almost a century later, in 1944, scientists identified DNA as the "particles of inheritance." The race was then on to determine its structure. Eventually, the research team of James Watson and Francis Crick solved this scientific puzzle in 1953. The structure that Crick and Watson conceived was a **double helix**. An **X-ray** image captured by English chemist Rosalind Franklin revealed that DNA was helical. Sadly, Franklin died before receiving full credit for her critical contribution. Scientists had proposed many other possible configurations, including a triple helix. However, only the double helix model fit the evidence of base-pairing provided by Erwin Chargoff.

Chapter 2: Blueprint of the Century: DNA

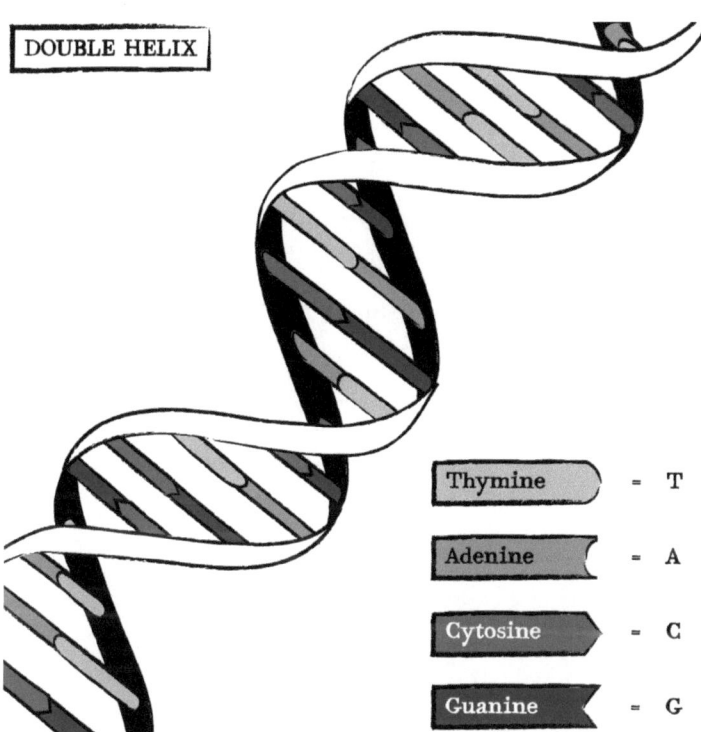

DOUBLE HELIX

Thymine = T
Adenine = A
Cytosine = C
Guanine = G

Cocktail Fodder

If you could type 60 words per minute, eight hours a day, it would take approximately 50 years to type the human genome.

The last sentence of Watson and Crick's *Nature* paper that described the double helix structure, made a critical point: "It has not escaped our notice that the specific pairing we have postulated immediately suggests a possible copying mechanism for the genetic material." This suggestion would have profound implications for the yet unrealized field of biotechnology.

Chapter 2: Blueprint of the Century: DNA

Cocktail Fodder

If unwound and linked together, the strands of DNA in each of your cells would be six feet long.

Life's Building Blocks

The full name for **DNA** is **D**eoxyribo**N**ucleic **A**cid. It's a long molecule classified as a nucleic acid because it is made up of repeating subunits called nucleotides. Each of the four DNA nucleotides consists of three parts: a sugar, a phosphate group, and a base. The prefix "deoxy" means "without oxygen." It refers to the lack of an oxygen molecule at a particular spot in the sugar molecule. "Ribo" stands for ribose, which is the type of sugar.

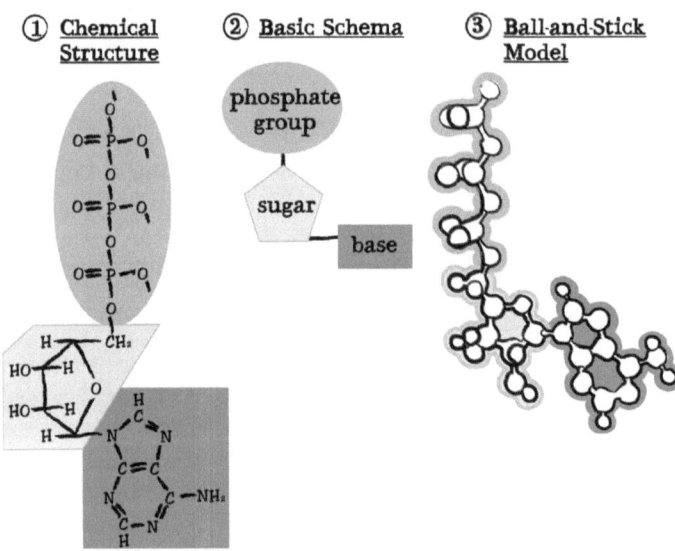

NUCLEOTIDE

A nucleotide is a repeating subunit of DNA. Each nucleotide is composed of a phosphate group, a sugar, and a base. Here are three different ways of viewing nucleotides:

① Chemical Structure   ② Basic Schema   ③ Ball-and-Stick Model

Looking at the *chemical structure* above notice the following:

Chapter 2: Blueprint of the Century: DNA

1. The sugar molecule called deoxyribose (the "D" in DNA) has five carbon atoms that form a ring structure.
2. The phosphate molecule is attached to the fifth carbon atom in the sugar ring which gives DNA weak acidic properties (the "A" in DNA).
3. The base is attached to the first carbon atom in the sugar ring.

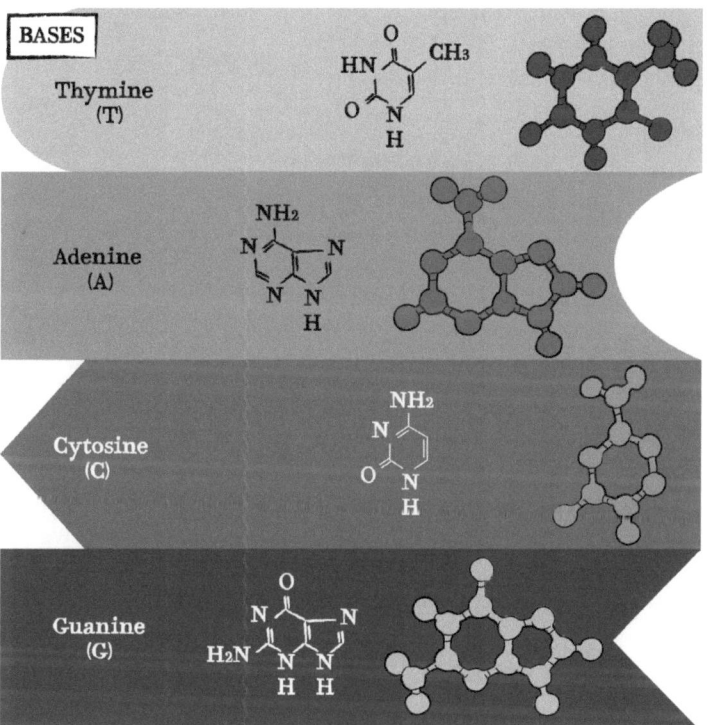

All DNA, found in all living things, use the same four bases called **thymine** (T), **adenine** (A), **cytosine** (C), **guanine** (G) and the same sugar and phosphate molecules. Incredibly, the nucleotide building blocks for the genetic material of all forms of life are identical.

Chapter 2: Blueprint of the Century: DNA

Cocktail Fodder

Where do cells get the nucleotide building blocks to make DNA? The food we eat. Animal and plant cells contain DNA made of the same constituent parts as ours! The DNA you consume in your salad or salmon is broken down into nucleotides that we reuse. Our cells can also synthesize the four nucleotides **de novo**, meaning from the beginning, using atoms from proteins, sugars, and fats that we consume.

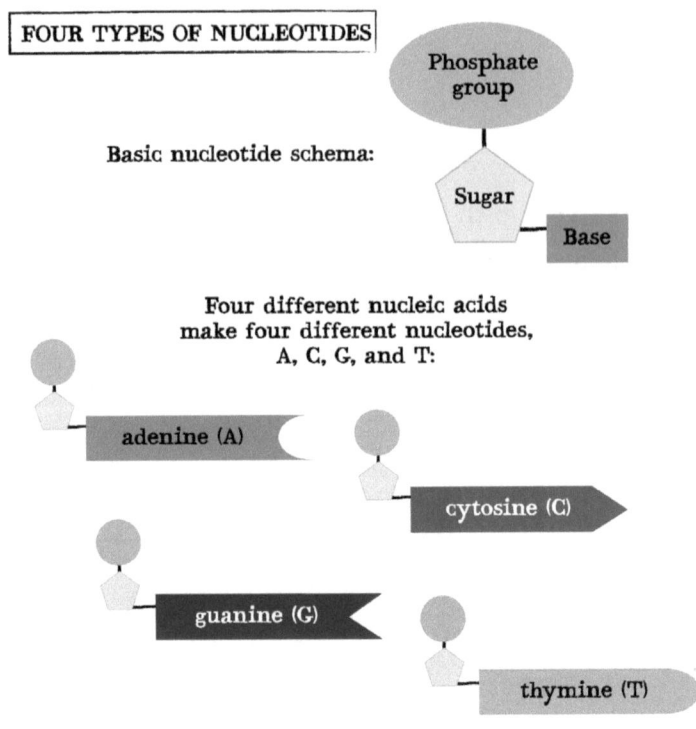

Putting It All Together

Nucleotides are linked to form a long chain. During DNA synthesis, a cellular enzyme called **DNA polymerase** links the phosphate group of one nucleotide to the sugar group of another. This forms what is often referred to as the sugar-phosphate backbone.

Chapter 2: Blueprint of the Century: DNA

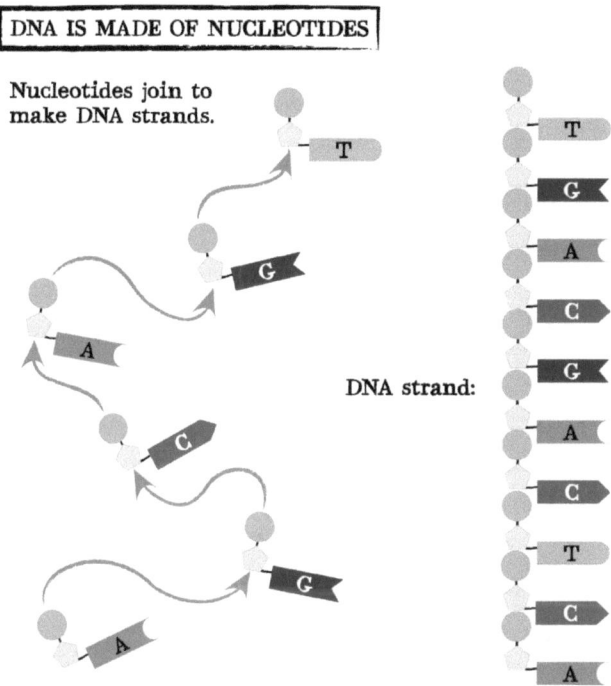

DNA IS MADE OF NUCLEOTIDES

Nucleotides join to make DNA strands.

DNA strand:

How do nucleotide chains form a double helix structure? An early experiment by Chargoff revealed some striking patterns. He noticed that within a DNA molecule, the number of Cs is always the same as the number of Gs. Similarly, the number of Ts is always exactly the same as the number of As. This led to the conclusion that DNA is composed of paired strands: Cs on one strand are matched to Gs on the other, and As on one strand are matched to Ts on the other. The pairing of Cs to Gs and As to Ts between strands is accomplished by chemical bonds. Their geometry, determined by the particular shapes of the nucleotides themselves, gives DNA its iconic shape.

Chapter 2: Blueprint of the Century: DNA

Cocktail Fodder

The largest known human gene—dystrophin—is made up of around 2.4 million bases. Dystrophin is required for proper muscle contraction.

Cocktail Fodder

Eight percent of human DNA is actually made of ancient viruses that once infected humans!

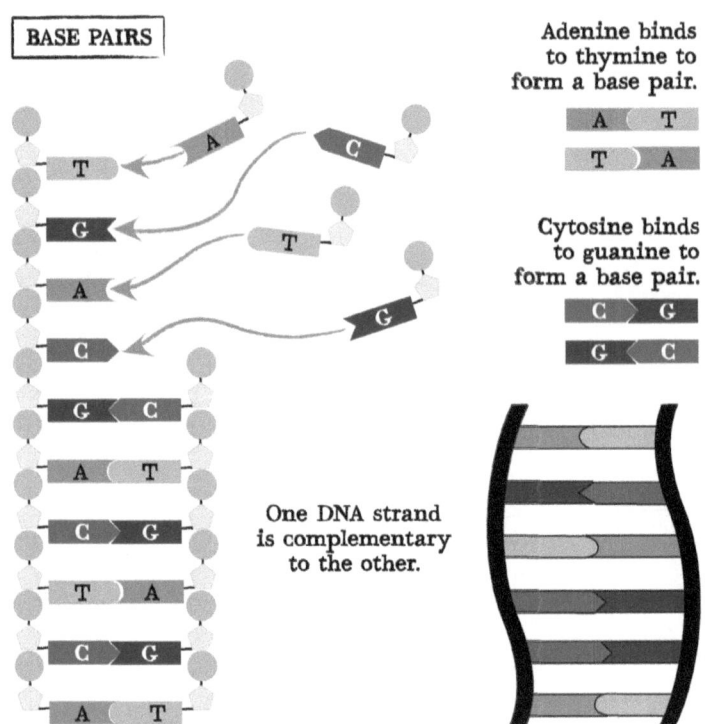

All organisms have DNA with the **same** four nucleotide bases. However, the bases are ordered **differently** in each. Their order is the **DNA sequence**. Every organism, from yeast to sugar cane to giraffes, has a unique DNA sequence.

The sugar-phosphate bonds that **form the molecule's** backbone are very strong. They give **the genetic information** in DNA great stability. This strength **makes** it possible to retrieve DNA from some fossils **and** explains why law enforcement personnel can collect DNA forensic evidence from a crime scene days, weeks, or **even months after** the event.

How DNA Replicates

Recall that from the start, Watson and Crick, believed the double helix model suggested a copying mechanism for genetic material. A mere five years after Watson and Crick published their seminal paper, American bio-chemist Arthur Kornberg verified how DNA replicates. His discovery of the enzyme that facilitates the process, **DNA polymerase**, garnered Kornberg a Nobel Prize in Physiology or Medicine in 1959.

Although the chemical bonds that connect adjacent nucleotides are very stable, the bonds holding complementary base pairs together are relatively weak. This makes the first step in DNA replication, *strand separation*, easily accomplished by an enzyme called **DNA helicase**.

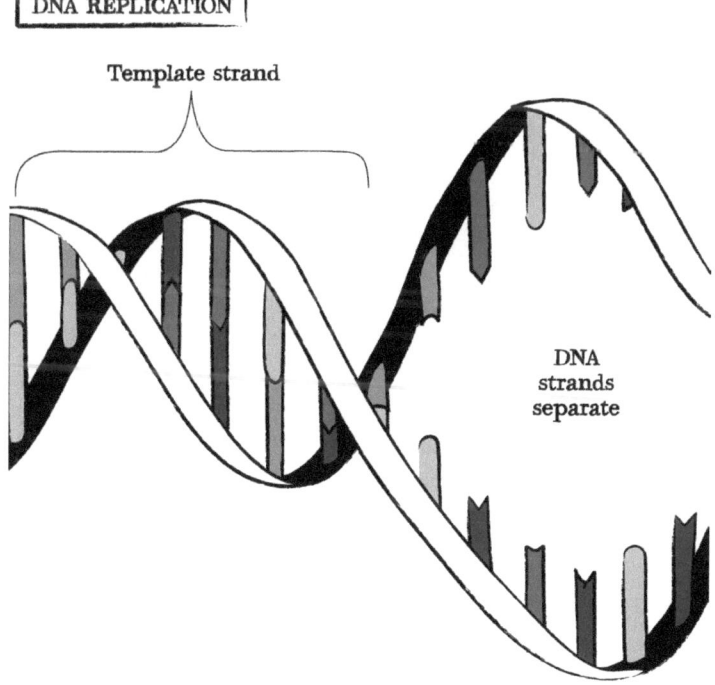

Once the two strands are separated, each one can act as a template for a new complementary strand of DNA. The appropriate free-floating base matches up to its appropriate exposed base on the separated DNA strand. A molecule called DNA polymerase bonds the newly aligned base-pairs, which creates the new complementary strand. Working one base at a time, DNA polymerase moves along each separated strand, using it as a template to produce its complementary strand until the chain is complete and the DNA is fully replicated.

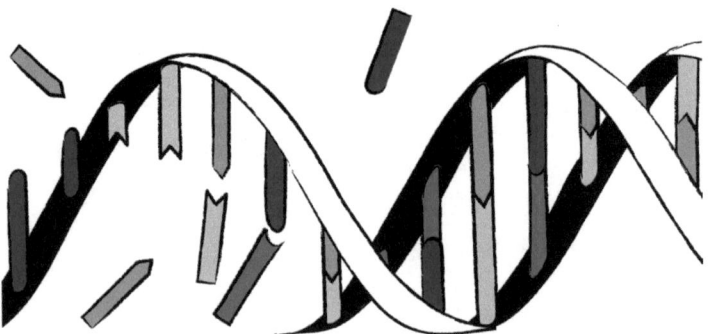

Complementary nucleotides fill in along template strands.

When it's all over, there are two new DNA molecules, each identical to the original.

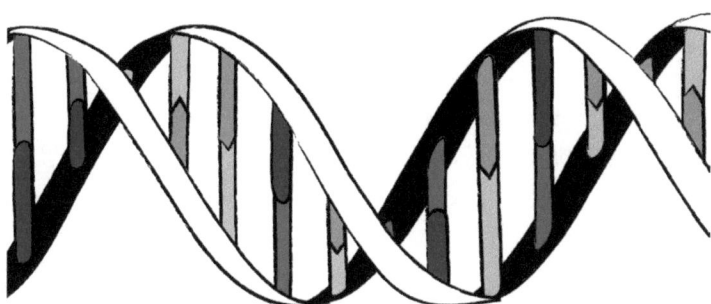

Two identical copies of original double helix.

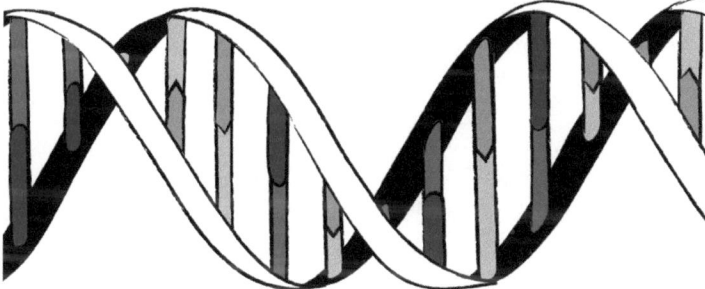

GOING FURTHER: Isolating DNA

To work with DNA, scientists isolate and extract genomic DNA from living cells. The process varies, but generally relies on three basic steps:

- *Lysis.* Cells are broken open, or lysed, in a detergent-containing solution. The cell membrane is made of lipids, which are fatty, "greasy" substances. That means the lipids dissolve in detergent, the way bacon grease comes off of a frying pan in the kitchen sink. When the cells have been broken open and their contents released, what remains is a **cell lysate**.

- *Separation from protein.* The detergent used in breaking the cell membrane serves a dual purpose. It also **denatures**, or unfolds, the proteins to which the DNA is attached. Denaturing helps separate the DNA. Most lysis buffers also contain **proteases**. Proteases are enzymes that break down, or **cleave**, proteins. These steps, denaturing and cleaving proteins, are necessary because scientists need protein-free DNA for their research.

- *Precipitation.* Next, the DNA is **precipitated** in ethanol. Adding ethanol causes the DNA to form into a white solid which can be easily isolated. The sample tube is then centrifuged, leaving a pure DNA pellet at the bottom. This **pellet** can be dissolved in a small volume of buffered water, stored in the freezer, and used for later applications. In an appropriately-buffered solution, frozen DNA is stable for several decades.

Organization of DNA

The genome of an organism is the full complement of its DNA. Each cell of the organism contains an identical copy DNA. These are organized into chromosomes, little packages of DNA. Chromosomes are highly condensed DNA wrapped around a protein called a **histone**. "Unwrapped," a chromosome looks simply like a long strand of DNA. Along this strand are many genes. A gene is a segment of DNA that codes for a protein. For example, the insulin gene, on chromosome eleven, provides the "instructions" to make insulin. Genes vary in size, from a few hundred to several thousand nucleotides long.

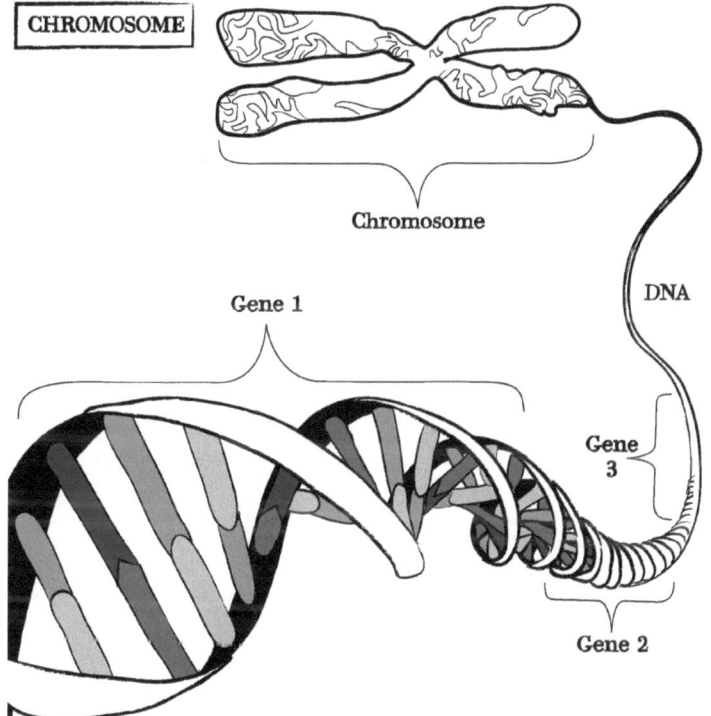

Cocktail Fodder

Orbiting our Earth is a memory device that is known as 'Immortal Drive'. The device is inside the International Space Station and it contains digitalized DNA sequence of Lance Armstrong, Stephen Colbert, Stephen Hawking and others. It is actually an attempt to preserve human race in event of a global catastrophe.

Different organisms have different numbers of chromosomes. Human cells, for example, have 23 pairs of chromosomes or 46 in all except in the germ line cells, also known as reproductive cells. Each egg and sperm has 23 chromosomes, a single copy of each. When a sperm cell fertilizes an egg cell, the chromosome count comes back up to 46. This means that everyone has two different copies of each chromosome; one from dad, one from mom.

Cocktail Fodder

If you put all the DNA molecules in your body end to end, your DNA would reach from the earth to the sun and back over 600 times!

Therefore, each of us also have two different copies of each gene. This idea gets more attention in Chapter 4, which covers genetic variation and precision medicine.

The information contained in DNA is read in groups of three bases, known as **codons**. Picture an organism's genome as a book. The nucleotides are the letters, the codons are the words, the genes are the sentences, and the chromosomes are chapters. Individual words and sentences have limited meaning; it's the collection of chapters as a whole from start to finish that tells the story.

INDUSTRY NOTE: Research Support Company

A **research support company** doesn't actively engage in drug discovery research. Instead, it supplies the tools and technologies required. For example, research support companies sell the DNA extraction kits used in the process described above. The same company might also provide enzymes needed to manipulate DNA in the lab, kits for making copies of specific gene sequences, and even the very sophisticated equipment for analyzing the human genome, DNA sequencing machines.

In this chapter, we've looked closely at the structure and organization of DNA. In the next, we examine how cells convert the information stored in DNA into the proteins that enable cells to function.

CHAPTER 3

Gene Expression: From Gene to Protein

The conversion of genetic information into a protein is called **gene expression**. It's a multistep process in which cells build new proteins. Proteins are made of **amino acids**. There are 20 amino acids and it is the order of these amino acids that determine the type of protein made. The amino acids' order is determined by the gene's DNA sequence. Amino acid chains are determined by two processes: **transcription** and **translation**. Let's take a closer look.

Step One: Transcription

During transcription, the original nucleotide sequence of the DNA code is rewritten into a molecule called **messenger RNA (mRNA)**. RNA molecules are very similar to DNA with three key differences:

- They have a ribose sugar rather than deoxyribose
- They're single-stranded, not double

- They use **uracil (U)** instead of thymine as one of their bases

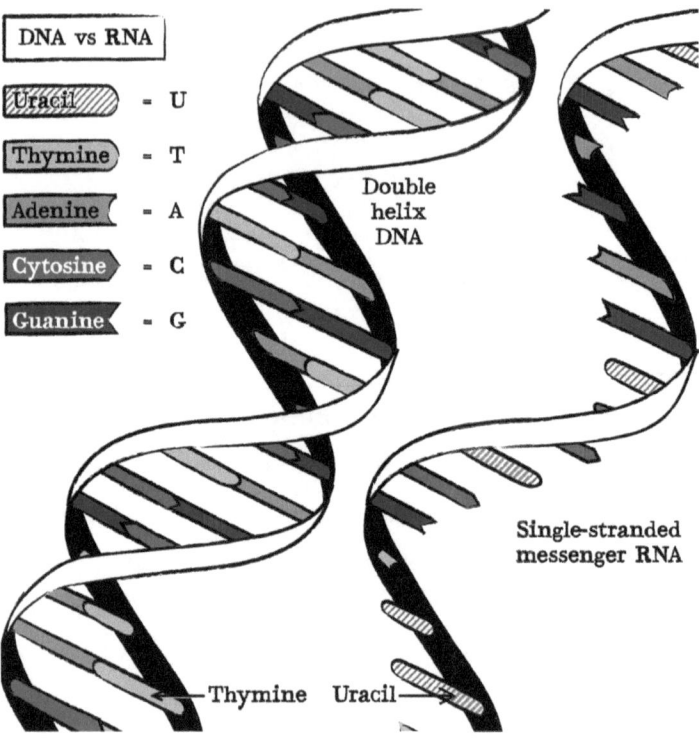

Each group of three nucleotides (also called a **codon**) in RNA specifies a particular **amino acid**. Sixty-four potential codons correspond to all possible combinations of the four bases A, C, G, and U. Because there are only 20 amino acids, most amino acids are coded for by multiple codons. In the 1960s, scientists labored to unravel the genetic code, that is, determining what combinations of bases represent which amino acids. By 1966, a team led by Marshall Nirenberg at the National Institutes of Health had figured it out. Nirenberg received the Nobel Prize for Physiology or Medicine in 1968 in recognition of this accomplishment. His results have been "codified" in the following table.

Codon Table

First Position	Second Position: Uracil		Second Position: Cytosine		Second Position: Adenine		Second Position: Guanine		Third Position
	code	amino acid	code	amino acid	code	amino acid	code	amino acid	
U	UUU	phe	UCU	ser	UAU	tyr	UGU	cys	U
U	UUC	phe	UCC	ser	UAC	tyr	UGC	cys	C
U	UUA	leu	UCA	ser	UAA	STOP	UGA	STOP	A
U	UUG	leu	UCG	ser	UAG	STOP	UGG	trp	G
C	CUU	leu	CCU	pro	CAU	his	CGU	arg	U
C	CUC	leu	CCC	pro	CAC	his	CGC	arg	C
C	CUA	leu	CCA	pro	CAA	gln	CGA	arg	A
C	CUG	leu	CCG	pro	CAG	gln	CGG	arg	G
A	AUU	ile	ACU	thr	AAU	asn	AGU	ser	U
A	AUC	ile	ACC	thr	AAC	asn	AGC	ser	C
A	AUA	ile	ACA	thr	AAA	lys	AGA	arg	A
A	AUG	met	ACG	thr	AAG	lys	AGG	arg	G
G	GUU	val	GCU	ala	GAU	asp	GGU	gly	U
G	GUC	val	GCC	ala	GAC	asp	GGC	gly	C
G	GUA	val	GCA	ala	GAA	glu	GGA	gly	A
G	GUG	val	GCG	ala	GAG	glu	GGG	gly	G

To use the chart, start on the left-hand side with the bold black letters. These letters represent the first position of the codon. The second position is at the top of the chart in bold. The third position of the codon is listed on the right-hand side, again in bold. For clarity, each three-letter codon is displayed to the left of its amino acid.

GOING FURTHER: Codons and Mutations

What explains the redundancy of multiple codons coding for one amino acid? It may protect against DNA mutations. For example, a mutation in the leucine codon could yield a mutated codon that *still* codes for leucine. Recently, however, scientists have learned that different codons for the same amino acid may be translated by the ribosome at different rates. This may influence the protein's ultimate shape and function.

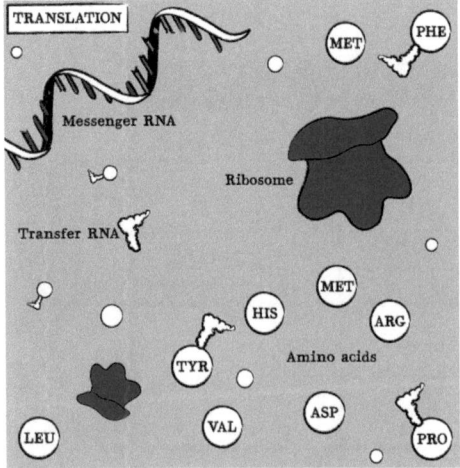
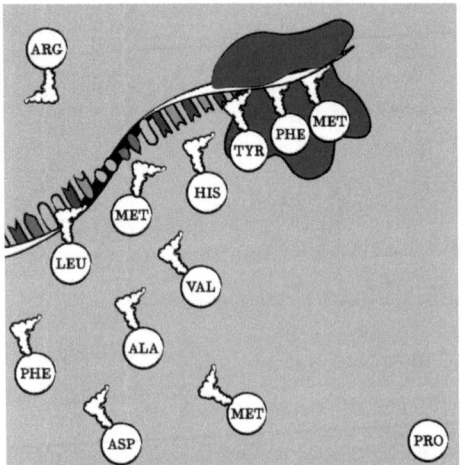
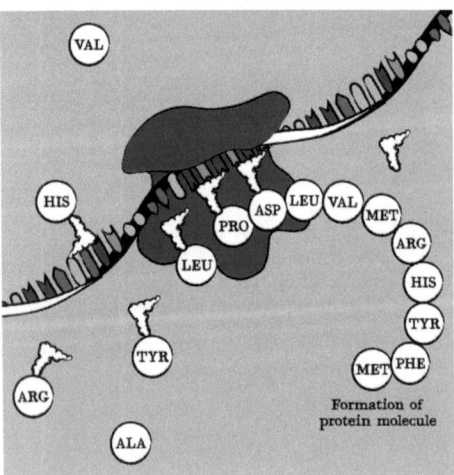

Formation of protein molecule

Step Two: Not Lost in Translation

The conversion of mRNA into a protein is called translation. It involves a second type of RNA, **transfer rna (tRNA)**. tRNA picks up amino acids in the order specified by the mRNA codons and transfers them to the ribosome. The ribosome forms a chemical bond (called a **peptide bond**) between the two amino acids. As the ribosome moves along the RNA message, it keeps connecting the specified amino acids to form a chain and the final protein molecule.

Nucleotide sequences also tell ribosomes where to begin and end translation. The **start codon** is AUG, which codes for methionine. As a result, the first amino acid of many proteins is methionine, though in some proteins, it's removed shortly after translation. There are three **stop codons**: UGA, UAA, and UAG. Stop codons don't code for amino acids, but are the last codon found in the protein sequence.

Proteins Take Shape

Ribosomes translate mRNA into a collection of amino acids like a string of beads, sometimes called a **polypeptide chain**. Each amino acid molecule has a common structure with three main parts: an amino group, an acidic carboxyl group, and an **R group**. Only the R group differs between amino acids. The order of amino acids in the chain determines the protein molecule's final structure.

Cocktail Fodder

Of the 20 different amino acids, scientists consider 10 essential. That's because we get them from our diet. Our bodies make the others. The essential amino acids vary for different organisms and even during different stages of life.

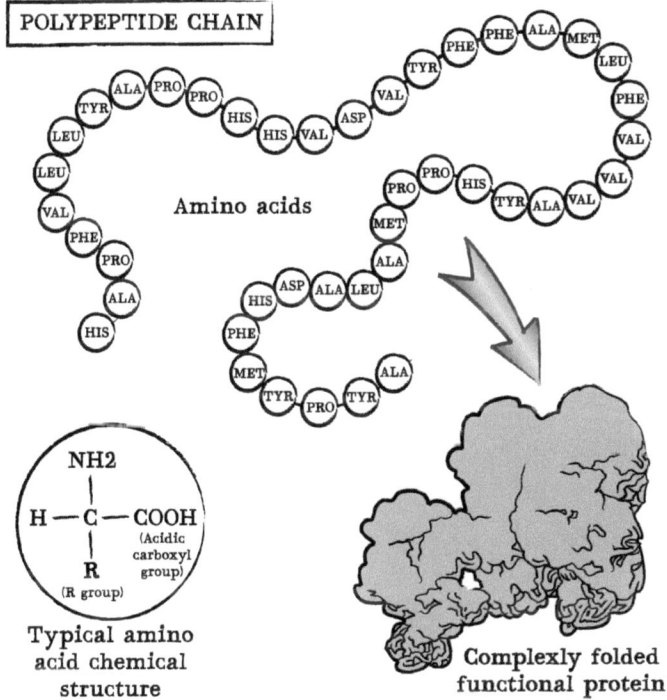

Some R groups attract one another; others repel one another. This interaction between points along the chain, as well as with the cell's aqueous environment, cause a protein to fold into a particular shape. Each unique sequence of amino acids forms a characteristic protein structure. The sequence makes the same structure every time it's translated. Incredibly, no two protein structures are the same. Even with only 20 amino acids, each protein molecule's structure is absolutely unique.

Once an amino acid chain folds to make a structure, it often interacts with other folded amino acid chains. In the picture below, four chains are interacting to form a structure different from each individual chain. In forming this elegant shape, a **tetramer**, the protein creates a new pocket that may function as an **active site.** Active sites are where proteins carry out their functions. The illustration depicts a hemoglobin tetramer. Its active site binds oxygen and transports it throughout the body.

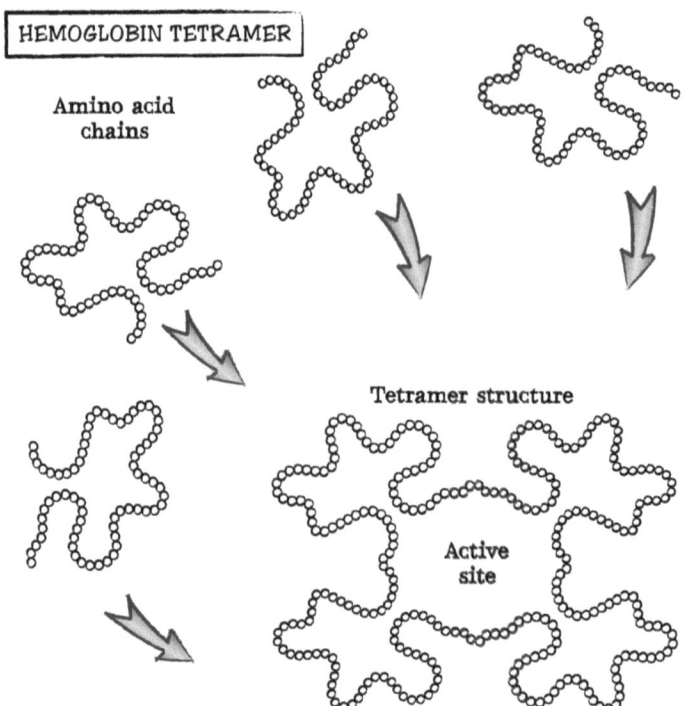

Chapter 3: Gene Expression: From Gene to Protein

GOING FURTHER: Protein Folding

Scientists are still trying to unravel the mysteries of how proteins fold. They've observed over one hundred types of protein folds using technologies such as **X-ray crystallography** and **nuclear magnetic resonance**. The field of **bioinformatics** uses computer modeling to predict the three-dimensional folds of new protein sequences compared to the structures of known proteins. Understanding protein architecture can help researchers identify new drug targets and design target-specific therapies.

Protein Therapeutics

The biopharma industry was founded on the idea that cells can be engineered to produce specific proteins for therapeutic purposes. Protein therapeutics encompass a wide range of proteins and therapeutic areas. Insulin was the first to become commercially available. Approved by the FDA in 1982 and manufactured by Genentech, insulin heralded the start of the biotech revolution in healthcare. It was quickly followed by human growth hormone, erythropoietin (which stimulates red blood cell production), immune-stimulating interferon, a range of enzyme-replacement therapies, and monoclonal antibodies. Over a hundred protein therapeutics are now on the market.

Protein therapeutics are also called large molecule drugs. This refers to the size of a protein therapeutic relative to a conventional drug. Proteins range from fifty to a thousand times bigger than a typical small molecule drug such as aspirin. These protein-based drugs may also be classified as biologic drugs. It's important to note, however, that the regulatory category of biologics includes

Tricky Terms

Meros means "part" in Greek. This is the origin of the root "mer." It's scientist-speak for a repeating unit. The prefix in front of **mer** varies, depending on how often the unit repeats. Hence: a single unit is a mono**mer**; two—a di**mer**; three—a tri**mer**; and four is a tetra**mer**. For large or variable numbers of repeat units, scientists use the term poly**mer**, "many parts."

other products such as vaccines and cell or gene therapies. *The Biopharmaceutical Primer: An Insider's Guide to Today's Advanced Therapies* dives deeper into these different types of biologic drugs.

Therapeutic proteins must be produced by cells. Scientists use genetic engineering to transfer a gene encoding the therapeutic protein to the production cell line. Thousands of liters of engineered cells are grown in order to produce enough therapeutics for a patient population. The whole process is called biomanufacturing. In contrast to biologic drugs, small molecule drugs such as aspirin are chemically synthesized. In Chapter 5 of this book, we explain the process of engineering cells to produce therapeutic proteins; in Chapter 9, we explain how this process is scaled up to produce large quantities of product.

When creating therapeutic proteins, scientists must consider how to keep them stable so they work safely and correctly in patients. Proteins are carefully formulated for maximum stability, but most require refrigeration. Another distinguishing factor of these drugs is how they are administered to patients. Because proteins break apart so easily in our digestive system, they must be provided by injection. In contrast, small molecule drugs are typically administered orally.

Peptide Therapeutics

The FDA defines a peptide therapeutic as a chain of amino acids containing fifty or fewer amino acids. It regulates them as small molecule drugs. Peptide therapeutics are similar in that they can be made using a peptide synthesis machine. This device links amino acids in a specified order. Peptides share key characteristics with large molecule drugs too. These include sensitivity to digestive enzymes, delivery by injection, and high specificity for their target.

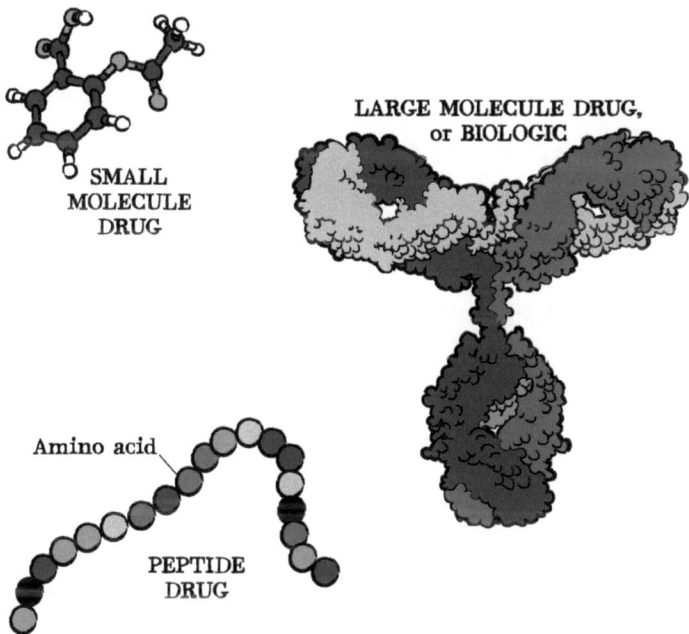

Examples of peptide drugs on the market include glucagon-like peptide-1 (GLP-1) and receptor activators. These medicines interact with a receptor on the surface of pancreatic beta cells and stimulate the release of insulin.

Cocktail Fodder

Friedrich Miescher discovered DNA in 1869, although scientists did not understand DNA was the genetic material in cells until 1943. Prior to that time, it was widely believed that proteins stored genetic information.

Protein Shape and Sickle Cell Disease

With an understanding of how gene sequence determines protein structure and function, let's look at the effects of genetic variation on cell function. Sickle cell anemia serves as an excellent illustration.

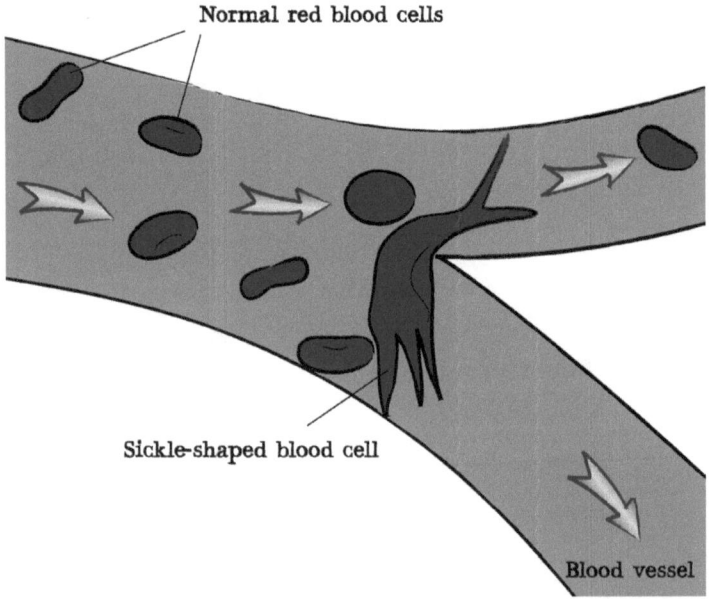

SICKLE CELL ANEMIA

Sickle cell anemia results from a single base **mutation** in the hemoglobin gene. The hemoglobin protein, coded for by the hemoglobin gene, consists of four protein subunits: two alpha and two beta. These form the functional hemoglobin tetramer. The mutation occurs in the beta subunits, which forms a "bump" in the protein's structure. This bump happens to fit into a notch in the beta subunit of the second hemoglobin molecule. The bump on that beta subunit fits into a notch on a third hemoglobin protein and so on, leading to the formation of rod-like hemoglobin chains.

Red blood cells with normal hemoglobin are flexible and assume the familiar round donut shape. However, the rod-like structures that result from the mutated gene mean that red blood cells with this variant become rigid and sickle-shaped. The malformed corpuscles can't travel through blood vessels as easily as the round normal ones. They get stuck, restricting blood supply and causing the complications of sickle cell disease. The body also removes the misshapen cells faster than the normal ones, which causes anemia.

Cocktail Fodder

Structural proteins are the most abundant class of proteins in nature. Collagen—the main component of connective tissues—is recognized as the most abundant mammalian protein.

GOING FURTHER: Transporting Oxygen

Hemoglobin transports oxygen in red blood cells from our lungs to the rest of us. The hemoglobin tetramer switches between two conformations, oxy and deoxy. In one, it binds oxygen; in the other, it releases oxygen. This switch is sensitive to **pH**, or blood acidity.

Blood in the lungs has a slightly higher pH than in the peripheral tissues. This causes hemoglobin to take on its oxygen-binding conformation. As blood leaves the lungs and enters, say muscle tissue, the pH is slightly lower, or more acidic. This small change in pH causes a conformational change in the hemoglobin protein. It then releases oxygen. Without the conformational switch, hemoglobin couldn't carry out its vital function.

Cocktail Fodder

Around 20% of our body is made up of proteins.

In the next chapter, we move from gene expression to genetic variation—differences in gene sequences among individuals. Scientific exploration of genetic variation has contributed invaluably to our understanding of disease and the development of new treatments and diagnostics.

CHAPTER 4
Genetic Variation

Our genomes—everyone's—are about 99.9% identical. That's why we can speak of something called "the human genome." However, its three *billion* base pairs mean that even a .1% difference in a DNA sequence can have big implications.

Genetic variation accounts for the differences we see among individuals' eye, hair, and skin color, for example. It also accounts for other differences we don't see such as blood type. Some variations are harmless; others seriously affect human health and account for our susceptibility to different diseases and response to their treatments.

All genetic differences arise from **mutations** in the DNA sequence. Even mutations themselves vary, as do their outcomes. Some common mutations include **substitutions**, **deletions**, and **insertion** errors.

Chapter 4: Genetic Variation

A substitution, or **point mutation**, results when one base is swapped out for another. If the DNA polymerase enzyme accidentally places a C where a G should be during replication, the substitution yields a different codon. Sometimes DNA polymerase skips over a base completely. That's a **deletion**. Or it adds an extra base to the sequence, which is an **insertion**.

Original sequence:
ATGACTGCA**T**GTTACGGT

Substitution mutation:
ATGACTGCA**C**GTTACGGT

Deletion mutation:
ATGACTGCA - GTTACGGT

Insertion mutation:
ATGACTGCA**GC**TGTTACGGT

Alterations as a result of errors by the DNA polymerase during replication are rare, but they do happen. Every time a cell divides, it copies three billion base pairs in just a *few hours*. So mistakes get made.

Fortunately, the polymerase also "proofreads" so it's able to correct most mistakes. Occasionally a mutation goes unrepaired and becomes permanent. In addition to replication errors, mutations also result from our environment. People smoke chemically-laden cigarettes, radiation from the sun bombards us daily and we get x-rays. It's a hazardous world out there!

Each and every one of our cells has mutations. This *sounds* alarming—but not all mutations are harmful. Mutations change how genes and their protein products function. Some changes, of course, *are* negative, and damage how an important protein works.

However, many mutations are "neutral" and don't change how a gene works at all. How is this possible? The Human Genome Project revealed that only about 1.5% of our genome codes for proteins. The rest of it is non-coding! So happily the odds are that most mutations take place in these swathes of non-coding regions and don't directly affect a protein's structure or function.

Non-coding DNA does play a role in regulating *when* the DNA that codes for proteins is used, so mutations in this region may affect how much of a particular protein gets made.

The other reason that mutations are so often neutral lies in the redundancy of the genetic code. As outlined in Chapter 3, most amino acids are coded for by more than one codon. Result: even if a mutation occurs in a protein-coding region, the protein gets made anyway.

Very rarely, mutations are **adaptive** and confer an advantage on the organism. For example, a slight change in an enzyme structure may make it more efficient. Constructive changes, *adaptations*, form the basis of Charles Darwin's theory of **Natural Selection** or "survival of the fittest." **Evolution** is the natural selection of beneficial changes.

Chapter 4: Genetic Variation

GOING FURTHER: Epigenetics

Epigenetics refers to chemical modifications to bases that don't change the DNA itself, but rather the extent to which the DNA is expressed. The most common **epigenetic modification** is DNA methylation. It adds the chemical group methyl (CH_3) to cytosine bases:

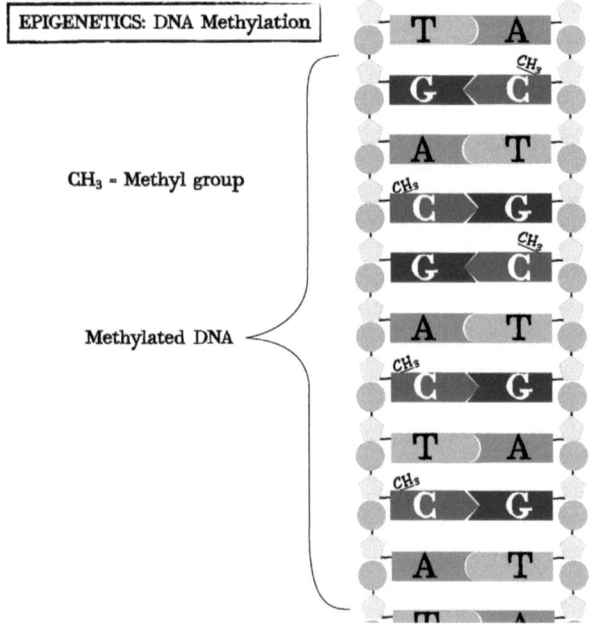

DNA methylation reduces gene expression by increasing the association of chromosomal DNA with the histone proteins it's wrapped around. It's a normal part of development, and an important reason that different genes are expressed differently in different tissues.

Another epigenetic modification is the direct chemical modification of histone proteins. This also impacts gene expression. Abnormal patterns of epigenetic modifications are linked to a daunting host of health conditions: cancer, autoimmunity, neurodegeneration, heart problems, stroke, depression. Histone modification makes an incredibly fruitful area for scientists developing drugs to address abnormal epigenetic modifications.

Some mutations go from one generation to another; some arise during a single life span. Most occur in **somatic** cells (body cells), not **gametes** (egg or sperm cells). Somatic mutations don't pass on to offspring. Those that *do* occur in sperm or egg cells, *germ line mutations*, get inherited. If the mutation is so severe that the offspring dies, clearly the mutation doesn't go on to the next generation. That's how harmful mutations eventually exit the gene pool.

Getting Snippy

Substitutions, in which one base is substituted for another, are more often known in the biotech industry as a **single nucleotide polymorphism (SNP**, pronounced "snip"). Scientists call small differences between different genomes **polymorphisms**. Generally, if a genetic variation shows up in more than one percent of the population, it's a polymorphism.

Single nucleotide polymorphisms are by far the most common. Further, they're most often the kind that contributes to our most obvious differences among us.

Other kinds of polymorphism also hugely affect genetic diversity. Examples include multiple base insertions or base deletions. In some cases, there are even differences in the number of copies of a particular gene—referred to as **copy number variation**.

Cocktail Fodder

Scientists estimate that, among individual people, there is a single difference per every 3,000 to 5,000 bases. That one base difference accounts for what makes each person unlike any other.

Tricky Terms

The term somatic derives from the Greek word soma meaning of the body. Thus somatic cell is the term used to refer to any cell of the body other than the gametes (egg and sperm). Gamete derives from the Greek words for husband (gametes) and wife (gamete). Gametes are also referred to as **germ cells**.

Genetic Basis of Disease: Monogenic Disease

There are two overall genetic categories of disease. In a **monogenic disease**, changes in one gene cause the illness. Examples include sickle cell anemia, cystic fibrosis, and Huntington's Disease.

Each of us bears two copies of each gene—one from mom, the other from dad. Each copy is an **allele**. A disease-associated gene is **recessive** if two copies of the abnormal allele are necessary for it to develop. People with one disease allele are **carriers**. They either don't have the disease or experience only mild symptoms. However, carriers pass on a disease *if* their partner is also a carrier.

A disease gene is **dominant** if the presence of only one copy of the abnormal gene can cause it.

From an evolutionary perspective, it makes sense that most serious monogenic diseases, such as cystic fibrosis, are recessive. Throughout most of human history, there was no treatment. People born with them most likely died before having children. Why then, do so many of

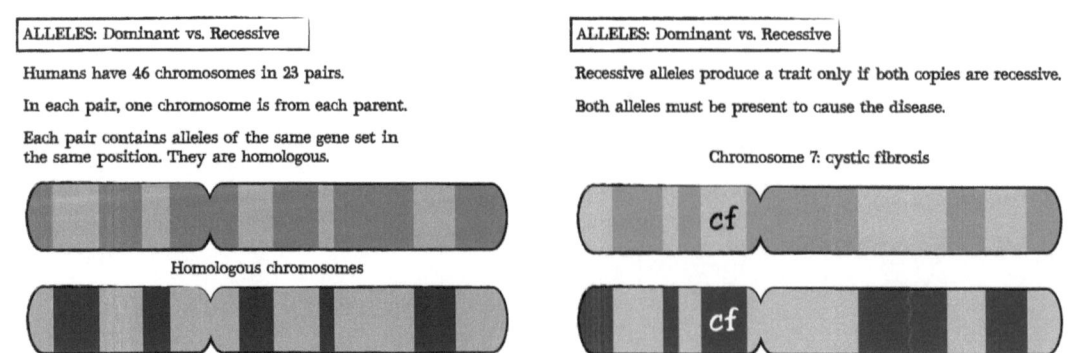

ALLELES: Dominant vs. Recessive

Humans have 46 chromosomes in 23 pairs.

In each pair, one chromosome is from each parent.

Each pair contains alleles of the same gene set in the same position. They are homologous.

Homologous chromosomes

ALLELES: Dominant vs. Recessive

Recessive alleles produce a trait only if both copies are recessive. Both alleles must be present to cause the disease.

Chromosome 7: cystic fibrosis

these genes persist in our gene pool? Their continued presence stems from the fact that having only one copy doesn't cause the disease. Take Huntington's disease, a deadly neurodegenerative disorder. The gene responsible for it is dominant. However, the disease manifests only in middle age, after the eventual sufferers have passed on the fatal gene.

However, some mutations confer an advantage. For example people with sickle cell trait are resistant to malaria. This deadly disease occurs in certain parts of the world, particularly Africa. Since they have protection against malaria, people with one copy of the sickle SNP have a survival advantage over those with two normal hemoglobin alleles in areas with a high incidence of malaria. The sickle cell SNP has therefore been *selected* for in most of Africa, where malaria has long threatened human health. The incidence of sickle cell disease among Africans is about four percent. Among people in the US of African-American descent, who aren't typically exposed to malaria-carrying mosquitoes, it's 0.25%. This is an example of the natural selection of a polymorphism in populations.

GOING FURTHER: Sickle Cell Anemia

> Why does sickle cell anemia occur much more often in some African populations than in Northern European ones, for example? This devastating disease occurs to people who inherit two hemoglobin alleles with the sickle cell SNP. There are two forms of sickle cell anemia—one much more serious than the other. A person with one copy of the SNP will have *sickle cell trait*, the much milder form.

Chapter 4: Genetic Variation

Cocktail Fodder

Blue-eyed people probably have a single, common ancestor, who had a genetic mutation between 6,000 and 10,000 years ago

Genetic Basis of Disease: Polygenic Disease

Polygenic diseases come from the interaction of many different genes. Polygenic diseases are more common than monogenic diseases. They include cancer, heart disease, Alzheimer's disease, diabetes, and Parkinson's disease. The illnesses often have **susceptibility genes** associated with them. These susceptibility genes increase the likelihood of the person developing the disease, but don't absolutely predict it. Instead, its occurrence depends on the rest of someone's genetic makeup as well as environmental factors.

For example, the apolipoprotein E gene (APOE) influences the development of late-onset Alzheimer's disease. People with different versions of the APOE gene have the following different risk profiles:

- Variant ε2 (APOE2): possibly lessens or delays Alzheimer's onset
- Variant ε3 (APOE3): is neutral
- Variant ε4 (APOE4): is associated with a significant increased risk of Alzheimer's

APOE proteins play a role in clearing the Alzheimer's-associated amyloid-beta plaque from our brains. Unfortunately, APOE4 performs this task less efficiently than the other variants of the gene. Evidence exists that APOE4 contributes to the breakdown of the blood-brain barrier that afflicts Alzheimer's sufferers, resulting in increased brain inflammation, another marker of Alzheimer's.

Remember that the presence of one or even two APOE4 alleles *isn't* diagnostic. Instead, large population studies suggest that individuals with the alleles will more likely develop Alzheimer's. Thankfully, not all do. A better understanding of APOE4's role in Alzheimer's onset will hopefully lead to the development of an entirely new class of drug.

Two of the best-known susceptibility genes are BRCA1 and BRCA2. BRCA stands for "BReast CAncer susceptibility gene." Someone with a mutation in these genes has an approximately *60%* increased risk of breast cancer and an approximately *40%* increased risk of ovarian cancer.

How does this mutation amplify risk without actually causing the disease? The dodgy genes code for proteins that are involved in DNA repair, especially in breast and ovarian tissue. Rather, the mutation raises the chances that someone won't be able to repair other mutations that do lead directly to cancer.

BIOPHARMA INNOVATION: PARP1 Inhibitors

Triple-negative breast cancers are HER2- and also lack estrogen and progesterone receptors. It doesn't respond to treatment with Herceptin or drugs that block hormones. Without receptor drug targets, this subtype poses formidable challenges. So far, no targeted therapeutics exist.

However, a new class of drug known as PARP1 inhibitors gives hope to people with triple-negative breast cancer. PARP1 is a second type of DNA repair protein. By inhibiting the PARP1 pathway, severe DNA damage triggers **apoptosis**, otherwise known as cell suicide. When a cell is cancerous, apoptosis is a very good outcome.

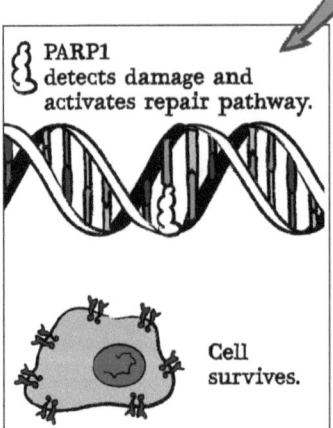

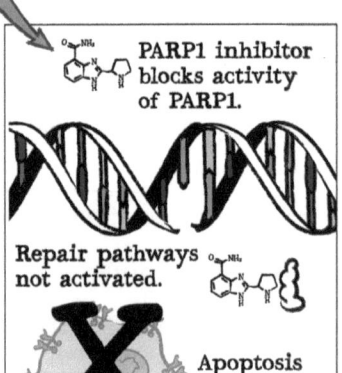

Cancer: A Polygenic Disease

Some cancers are traced to defects in one gene. However, most are thought to be polygenic; they develop only after mutations in different genes accumulate in a single cell. Mutations that lead to cancer are those that produce proteins regulating cell growth and division. The single genetically-mangled cell goes into uncontrolled cell growth. The runaway growth puts extreme stress on the affected tissue. Cancer cells no longer serve the tissue it's growing in. But they still occupy space and absorb nutrients, further preventing the healthy parts of the tissue from functioning correctly.

Different types of genes are likely to be mutated in cancer:

- **Proto-oncogenes**: These genes normally produce proteins that play an important role in healthy cellular growth and development. Mutations convert them to **oncogenes**. They either produce too much of the growth-inducing proteins or produce overactive proteins that signal cells to grow when they shouldn't.

- **Tumor suppressor genes**: These produce proteins that normally inhibit cell division, acting as an "off" switch for proto-oncogenes. Mutations cause them to malfunction and fail to shut down cell division correctly.

- **DNA-repair genes**: These produce proteins that repair mutations in other genes. So while a mutated

Cocktail Fodder

Some women have a genetic mutation that makes them tetrachromatic, which causes their eyes to have four different types of cone cells, enabling them to see 100 million different colors compared to the roughly one million colors most of us can see.

Cocktail Fodder

You have 98% of your DNA in common with a chimpanzee.

Chapter 4: Genetic Variation

Cocktail Fodder:

Even though most cancer cells bear many different mutations, targeted therapies such as Herceptin that inhibit a specific mutated oncogenic protein can be highly effective. Researchers believe such therapies work well because they rely more heavily on specific oncogenic mutations. This phenomenon has been termed "oncogene addiction."

DNA-repair gene on its own doesn't directly cause cancer, it can make the cell more likely to incur mutations because it's not repairing DNA.

Pharmacogenomics

The understanding that most diseases have genetic influences has led to a new area of medicine called **pharmacogenomics** or **precision medicine**. It uses information in a patient's genome to design the best treatment plan for that patient. Precision medicine takes into account potential disease genes, but also those that might affect someone's response to a drug.

Pharmacogenomics could enhance medicine in several ways, including predicting drug effectiveness. It may enable doctors to select the drug most likely to benefit a patient or, conversely, select the patients most likely to respond to a given drug.

Precision medicine also helps predict drug safety. Approximately 100,000 people die annually from adverse drug reactions, one of the leading causes of death in the US. Genetic differences, particularly in genes that metabolize drugs, mean that some people respond badly to certain medicines. Pharmacogenomics will help doctors avoid prescribing drugs that endanger a patient. Similarly, scientists can use drug effectiveness and safety data to develop an appropriate dosing schedule for the patient. These efforts also help control drug costs.

Benefits of Pharmacogenomics

Designing drugs to target **genotypes** will allow companies to fine-tune drug development from the beginning. Clinical trials will be improved and streamlined by selecting patients for trials by genomic signature, resulting in shorter, less expensive trials. Drugs that fail trials because of ineffectiveness could be reevaluated in genomically-selected patient populations. The lung cancer drug Iressa, for instance, didn't prove effective until scientists realized it only benefits patients with a mutation in the EGFR gene.

Biotechnology companies may be able to reduce drug costs by targeting only appropriate genotypes, selecting applicable clinical trial protocols, and focusing marketing on relevant groups. Finally, screening for genes that predispose to disease may provide more specific information on what we can do to protect ourselves from disease in the first place. This kind of screening is already underway for some diseases, such as breast cancer.

Precision Medicine in Practice

Precision medicine is happening now. In this section, we explore two examples involving cancer-related genes.

Genentech designed Herceptin to work on breast cancer cells that produce very high levels of **HER2** protein, a growth factor **receptor** involved in promoting cell division, called HER2+ cells.

Chapter 4: Genetic Variation

Cocktail Fodder

Cancer is an ancient disease. Evidence of cancer has been observed in human mummies from ancient Egypt; 3,000-yr old writings reference the disease.

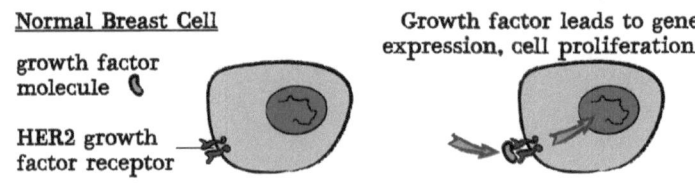

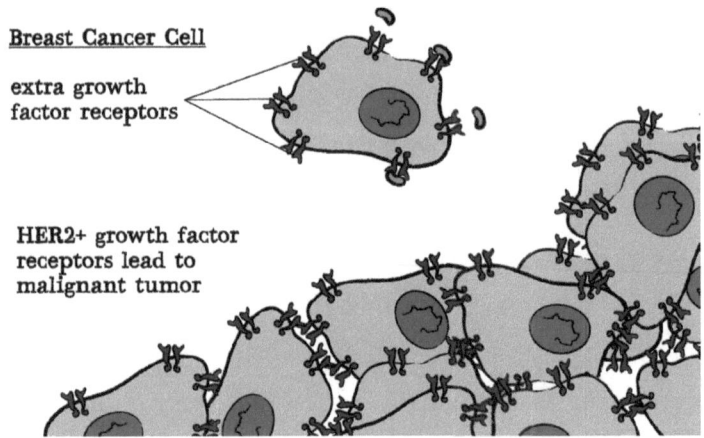

Herceptin, a monoclonal antibody, won't work in patients with HER2- tumors. So patients can be screened to predict the likelihood that Herceptin will help.

More recently, scientists have discovered another gene can help predict Herceptin's effectiveness. **PTEN** is the second most common tumor suppressor gene mutated in human cancer, after **p53**. Normally, **tumor suppressor genes** protect cells from becoming cancerous. When PTEN is mutated, this function fails and cells are more likely to become cancerous. Mutation of PTEN in HER2+ tumors seriously decreases Herceptin's effectiveness. Therefore, patients who are HER2+, and without mutations in PTEN (PTEN+), will probably respond well to Herceptin. Even

Chapter 4: Genetic Variation

better, it's very likely that other genes impact the effectiveness of Herceptin and other cancer drugs.

PRECISION MEDICINE: Finding the Right Therapeutic

HER2-
will not respond
to Herceptin

HER2+
some will respond
to Herceptin

HER2+ / PTEN+
most will respond
to Herceptin

Cocktail Fodder

DNA testing in criminal cases began back in 1985, and was first used to convict a criminal, Florida rapist Tommie Lee Andrews, in 1987.

Chapter 4: Genetic Variation

BIOPHARMA INNOVATION: Companion Diagnostics

Herceptin and other targeted therapeutics can be extremely effective, but only in the genetic subpopulations they are designed to treat. Therefore, physicians need a way to identify the patients most likely to respond. Enter **companion diagnostics,** which identify patients who carry the mutation the therapy was designed for. Several companion diagnostics are already on the market to identify HER2+ patients. Because HER2+ breast cancer is caused by a **gene amplification** event, there are extra copies of the gene that codes for the HER2 receptor. HER2+ breast cancer can be identified either by looking for elevated HER2 levels or for signs of an amplified HER2 gene.

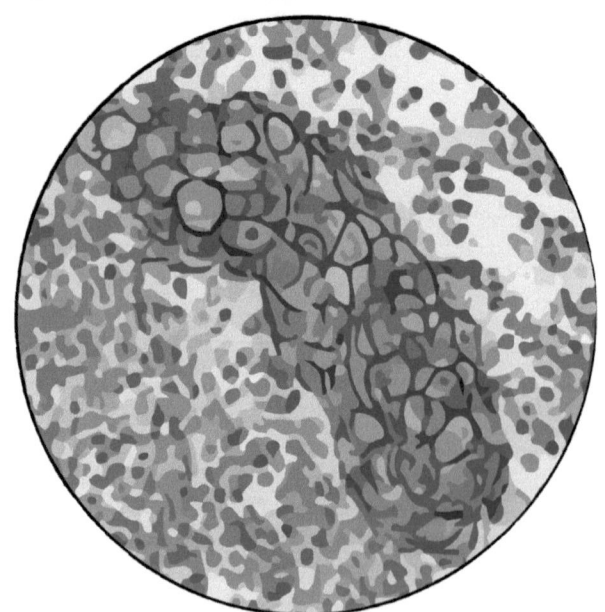

PRECISION MEDICINE: Herceptin Companion Diagnostics

HercepTest measures HER2 protein levels.

Another application of personalized medicine involves a large family of genes called **cytochrome P450**, or **CYPs**. They code for proteins that detoxify chemicals in the liver. They also metabolize many prescription drugs by breaking down the active drug into inactive products.

SNPs often occur in CYP genes of which there are many, including CYP2D6 and CYP3A4. This sometimes affects the ability of the CYP proteins to metabolize drugs. A SNP that decreases CYP activity will slow down drug metabolism, and cause unexpectedly high drug levels in the bloodstream for longer than expected. This can have very serious implications for drug effectiveness and safety. Many serious adverse reactions to anti-depressants, for instance, have been attributed to SNPs in CYP2D6. Knowing the CYP genotype of a patient is very important for designing an appropriate drug dosing schedule.

In this chapter, we've looked at genetic variation and its impact on medicine and disease. Next, we turn to the tools scientists use to better understand the human genome and to characterize disease-specific mutations: the field of **genomics**.

Cocktail Fodder

Some naturally occurring chemicals can decrease the ability of CYPs to metabolize drugs. For instance, grapefruits contain a chemical that blocks the activity of CYP3A4, which is why many drug labels advise against taking certain medications after drinking grapefruit juice.

CHAPTER 5
Deciphering Genetic Variation

As we saw in Chapter 1, most diseases share some connection with the alphabet soup that is our DNA. However, the influence of genes on our health varies profoundly. They range from inherited disorders that hinge on one mutation, like sickle cell anemia, to those that the medical community believes partially arise from susceptibility genes—which affect a person's long-term health but don't cause an illness directly. These include Crohn's and Alzheimer's diseases. One of the most feared maladies of all, cancer, is understood to be **polygenic**—stemming from mutations in multiple genes. Our genetic profile might even impact our vulnerability to some infectious illnesses. Genetics influence malaria, tuberculosis, and HIV. Now, we turn our attention to the tools that scientists use to better understand the **genome** and the influence of mutations on human health.

Our Secret Code

First, what is a genome? The term simply means the entire sequence of an organism's DNA. This biological code contains all of the information needed to make a person (or daisy or giraffe) and take them from birth to the end of their days.

Cracking the Code: The Human Genome Project

In 2003—fifty years after James Watson and Francis Crick deciphered the structure of DNA—the first human genome sequence was published. This astounding effort determined the order of the base pairs that make up our genome. It also identified and mapped the locations of all of those genes on our chromosomes. The project took thirteen years, cost nearly three billion dollars, and required collaboration among molecular biologists and geneticists from the United States, the United Kingdom, France, Germany, Japan, and China. To obtain this biological blueprint, researchers pooled DNA samples from about 100 volunteers recruited from public advertisements near the labs where the sequencing was conducted to come up with a representative genome sequence. Completion of the **Human Genome Project** is considered one of the major scientific advances of the twentieth century. As Francis Collins, then director of the National Human Genome Research Institute, which coordinated the project, said: "It's a history book—a narrative of the journey of our species through time. It's a shop manual, with an incredibly detailed blueprint for building every human cell. And it's a transformative textbook of medicine, with insights that will give health care providers immense new power to treat, prevent, and cure disease."

The data from the Human Genome Project was made publicly available. It's enabled researchers to better identify specific genes as being associated with certain diseases and catalog their common variations. They've also characterized the much larger regions of the genome *with no* genes. These areas are called non-coding or **intergenic DNA.** We've since learned that swathes of this intergenic zone help determine when our bodies use the coding portions.

Researchers had sequenced *individual* genes prior to the Human Genome Project. Some of these efforts yielded important medical advances. For example, knowing the sequence of the insulin gene enabled the founders of Genentech to transfer that gene into bacterial cells to manufacture human insulin—the first biotech drug.

Biotech companies quickly recognized the value of the Human Genome Project data. They also realized they needed new technologies to make large-scale, full genome sequencing possible. Less time and money per genome means researchers can collect information from the genomes of multiple patients and identify more predictable patterns associated with diseases. We're now at that point. Thanks to **next generation sequencing (NGS)**, the cost of analyzing an individual human genome has plummeted from over a billion dollars to a thousand. The process takes

Tricky Terms

Genetics is the study of how traits travel from one generation to the next. For example, the BRCA1 gene confers an inherited risk of breast and ovarian cancer.

Genomics is the study of gene sequences, gene expression, and the interaction of genes with each other and with regulatory elements, or regions of DNA that control how often regions of the genome are used.

only 24 short hours. All signs indicate sequencing will continue to get cheaper and faster—pricing is expected to go from a thousand dollars to a measly hundred, and from 24 hours to just a few hours.

Pieces of You

There are different approaches to next generation sequencing. One of the most popular is "sequencing by synthesis." This involves determining the order of one strand of DNA by detecting what base gets incorporated at each position in the synthesis of the new strand. In human DNA, A always pairs with T and C always pairs with G. By recording the arrangement of bases in the newly-synthesized strand, the NGS machine determines the sequence of the template strand. We'll discuss the details of *how* the machine recognizes which base gets incorporated below. Let's examine how a technician prepares a genome for sequencing.

The first step consists of breaking down the patient's genome into manageably-sized fragments for the sequencing machine to process. Individual chromosomes ranges in length from 50,000,000 to 300,000,000 base pairs. NGS machines only process DNA strands a few hundred base pairs long. That means to sequence someone's entire genome, technicians must first prepare a **sequencing library**—a collection of DNA fragments, created from one strand of DNA.

These fragments come from either cutting the DNA to be sequenced with enzymes or sonication—applying sound energy to break it apart. A library usually derives from 30 copies to ensure high levels of accuracy. Remember every cell contains a copy of your genome. Consequently, one blood sample gives technicians access to millions of copies.

This stage of the process results in the creation of millions of fragments. Then, technicians add **adaptor sequences** to both ends of each piece. Adaptor sequences are short sections of DNA with a known sequence—no more than 20 to 40 base pairs long. One adaptor sequence attaches each DNA fragment to the **flow cell** (described next) in which the sequencing reaction occurs. The other is used to start the sequencing of each genetic chunk.

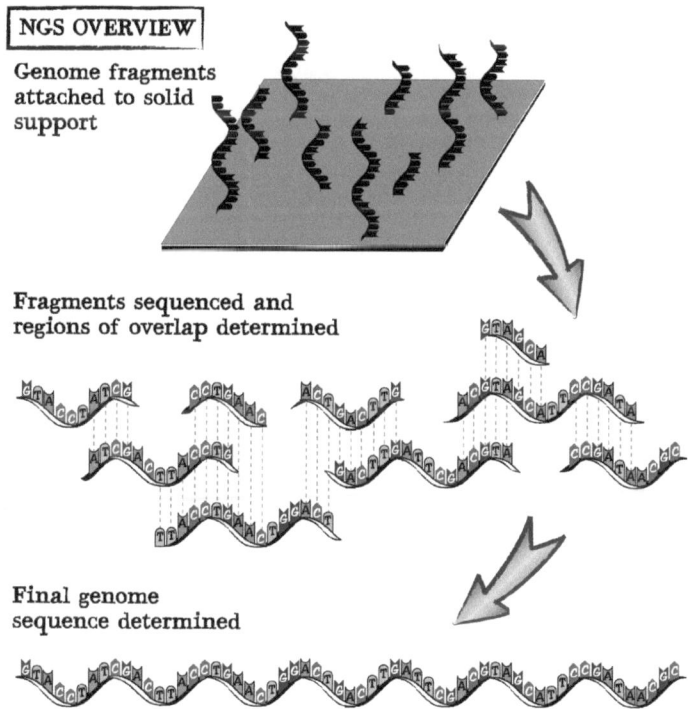

Sequencing by Synthesis: "Hi-Seq"

Let's take a closer look at one company's approach.

Illumina's (San Diego, CA) Hi-Seq machine uses a method known as *reversible dye terminator*. Simply put, Hi-Seq uses color coding to produce a *readout*. For each round of sequencing, the machine pumps four nucleotides over the DNA fragments attached to the flow cell. Each nucleotide has a specific fluorescent tag associated with it. For example, each A may be fluorescent green or each C, fluorescent yellow. The Hi-Seq machine also adds a DNA polymerase enzyme to connect the nucleotide bases. When this enzyme attaches a new nucleotide onto the strand being synthesized, the machine detects which color is incorporated and correlates it to the corresponding base.

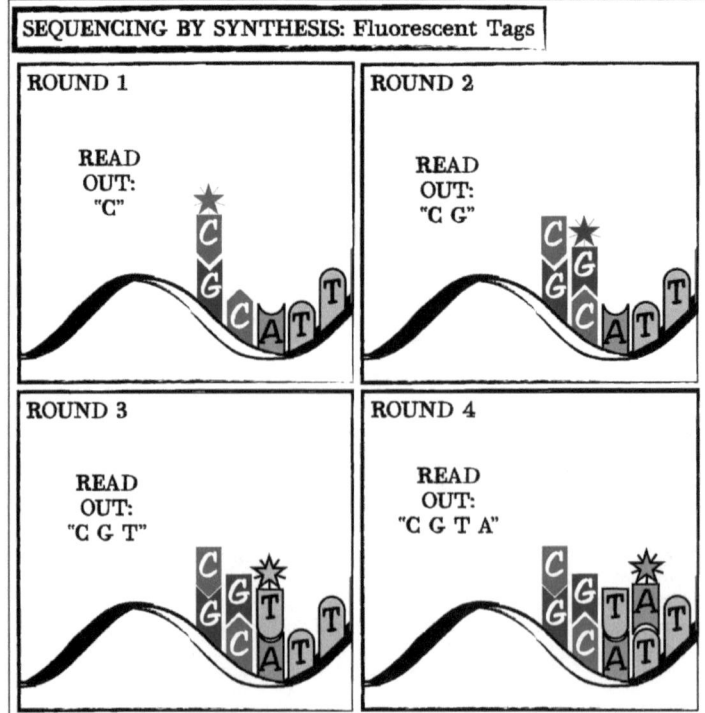

that contain overlapping sequences are patched together. This step relies on a computationally complex algorithm that compares the sequences of the hundreds of millions of DNA fragments to find the overlap.

Machines That Read: Third Generation

The advances in whole genome sequencing have been truly impressive thus far. However, a third generation of sequencing technology may revolutionize genomics yet again—with speedier, less expensive, and even smaller machines.

Third generation sequencing is sometimes referred to as "single molecule sequencing." Unlike NGS, it doesn't require the painstaking, step-by-step synthesis by which a machine infers the presence of particular bases. Third generation technologies instead directly "read" the bases.

Let's explore just one NGS technique. Single molecule sequencing or *nanopore sequencing* passes a single strand of DNA through a tiny pore in a synthetic membrane. Each pore is so miniscule that a strand can only go through one nucleotide at a time. Meanwhile, the machine simultaneously jolts the membrane with an electric current. As each nucleotide—A, C, T, or G—pops through, it disrupts the current in a characteristic way. The machine notes the nucleotide's distinctive commotion, and records its corresponding base.

GOING FURTHER: Flow Cells

A **flow cell** is a glass slide into which *nanowells*—miniscule compartments with a diameter of 10^{-9} m—are etched, using optical lithography—the same technology for manufacturing semiconductor chips. Each nanowell or "well" contains a short piece of synthetic DNA with a sequence that complements the adapter sequences attached to the fragments in the patient's DNA library. The technician then adds the DNA library to the flow cell and the patient's DNA fragments attach to the complementary sequences. Each fragment is then copied many times using **polymerase chain reaction (PCR)**, a standard procedure for replicating (making copies) a piece of DNA. Next, thousands of copies of each DNA fragment are attached to the surface of different nanowells. The machine then chemically removes one strand of each double-stranded fragment, leaving a single stranded copy of each for the sequencing reaction.

To recap: a technician prepared DNA libraries from 30 copies of the patient's genome. The sequencing machine then amplifies or increases the amount of each DNA fragment from the sequencing libraries thousands of times. The machine then determines the sequence of *each* of the hundreds of millions of DNA fragments. This is what the industry calls *"massively parallel sequencing."*

The flow cell also contains small channels through which buffers and the other **reagents** required for the sequencing can be pumped.

Once the library fragments are attached to the flow cell and amplified, the sequencing machine can begin its work.

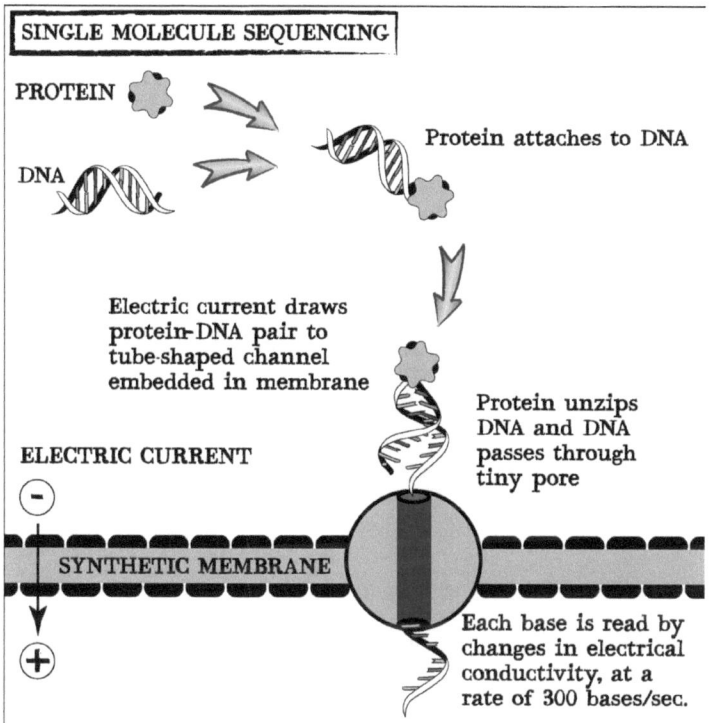

Single molecule sequencing hasn't yet achieved the time and cost efficiencies of sequencing by synthesis. However, it is anticipated that as third generation equipment improves, its efficiency and accuracy will surpass that of its forerunner. The approach holds several advantages over current practices. First, although deciphering an individual's genome still requires snippets of DNA, the pieces can now be much longer. They consist of thousands of bases rather than hundreds. This makes the data analysis— compiling someone's entire genome—less complex. Directly reading the DNA fragments is also less likely to introduce errors than sequencing by synthesis. This is because the polymerase used to synthesize new DNA can sometimes make mistakes.

Whole-genome sequencing is revolutionizing medicine. Cancer patients' medical choices no longer consist of surgery, chemotherapy, or radiation. Now, oncologists and other health care professionals are using genomics to further our understanding of how genes influence cancer. That information is already helping develop better testing and treatments specific to genetic subtypes of the disease. For example, the drug Zelboraf was designed especially for melanoma patients with a specific mutation that codes for an enzyme involved in cell growth. As the technology matures, it will open up applications in other areas of medicine—even its use as a routine point-of-care test to pinpoint mutations that increase the risk of disease.

Application: Deciphering a Killer

Currently, whole-genome sequencing mostly occurs in studies exploring the genetic basis of cancer. As a polygenic disease, several mutations typically influence a malignancy. Nailing down precisely which result in particular cancers enlarges researchers' understanding. In turn, that knowledge may pave the way for more precise, less debilitating treatments. When someone with cancer volunteers to have their entire genome read, they help scientists identify previously unknown genetic flaws. If a treatment exists, the volunteers either receive that treatment or enter a trial testing it if they wish. Universities, government researchers, and biopharmaceutical companies are currently compiling databases of mutation types and treatment outcomes in order to identify the best treatments for specific genetic subtypes of cancer.

Precision treatments feature a flip side: **companion diagnostics**. The kind of efficient third generation whole-genome sequencing may one day lead to a "universal companion diagnostic." Rather than checking if someone tests positive for Mutation X associated with cancer X that will likely respond to a specific drug, a universal companion diagnostic could potentially decode the person's full genome to probe thousands of mutations at once. Now that's modern medicine!

Easily Confused: Genome Sequencing vs. SNP Genotyping

A **single nucleotide polymorphism (SNP)** is a difference in one base pair between two DNA sequences. People sometimes misuse the term *DNA sequencing* to describe what's actually *SNP genotyping*.

- **DNA sequencing** determines *the order of every single base pair* in a given gene (gene sequencing) or an entire genome (whole genome sequencing).

- **SNP genotyping** identifies *single base changes* between two DNA strands—a given gene sequence and a reference sequence.

Tricky Terms

Exome Sequencing

What's an exome? If it sounds like part of our genome, you're right. It's the specific bit that codes for proteins—in other words, genes. It's possible to completely sequence our exome, which makes up only about two percent of the genome. In some cases, exome sequencing provides enough information to better understand a disease. However, scientists are realizing more and more that the intergenic regions of the genome play an important role in regulating when our bodies use the coding regions. In addition, because of course DNA is still present in the non-coding areas, disease-associated mutations also occur there.

A Chip Off the New Block: SNP Chips

It's often possible to learn valuable genetic information without full-genome sequencing. Using "SNP chips," researchers can quickly identify clinically relevant SNPs. Let's examine how the chips work and some of their applications.

A SNP chip is a tiny glass or silica chip with a microscopic checkerboard etched into it. Each square measures only 11 micrometers (10^{-6} meters) by 11 micrometers or so, just big enough to hold one single-stranded piece of DNA. Each strand represents one gene or gene variant (also known as an allele). This holds true regardless of whether or not the entire gene sequence is present. Typically, the process uses between 25 and 60 bases to uniquely represent one gene. High-powered software tracks the location of DNA sequences on the chip. The short, representative strings of DNA used in the process are called *probes*. A single SNP chip holds from ten to a few million DNA probes.

Chapter 5: Deciphering Genetic Variation

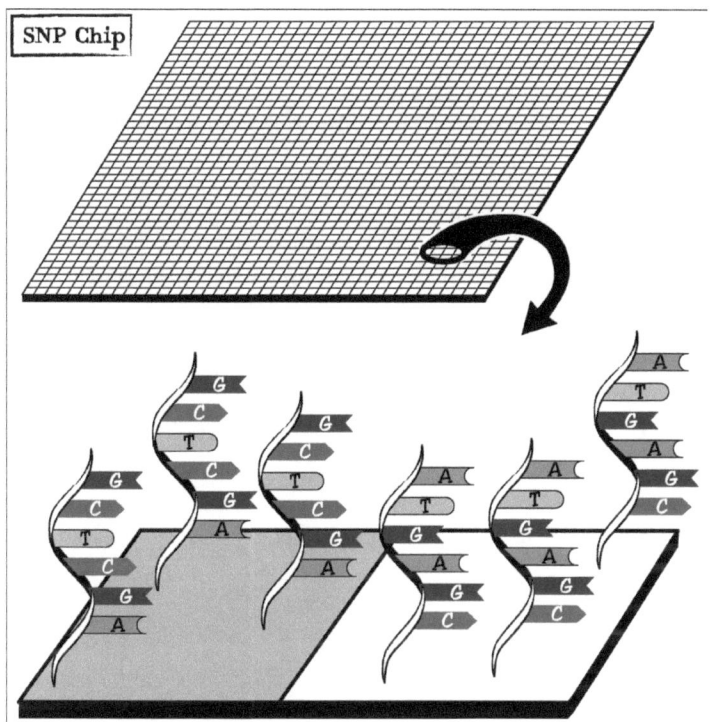

The goal here is to identify SNPs in a patient's DNA. A technician washes the patient's fluorescently labeled DNA samples over the chip to reveal which probes the patient's genes hybridizes or sticks. "Sticky" squares indicate a match.

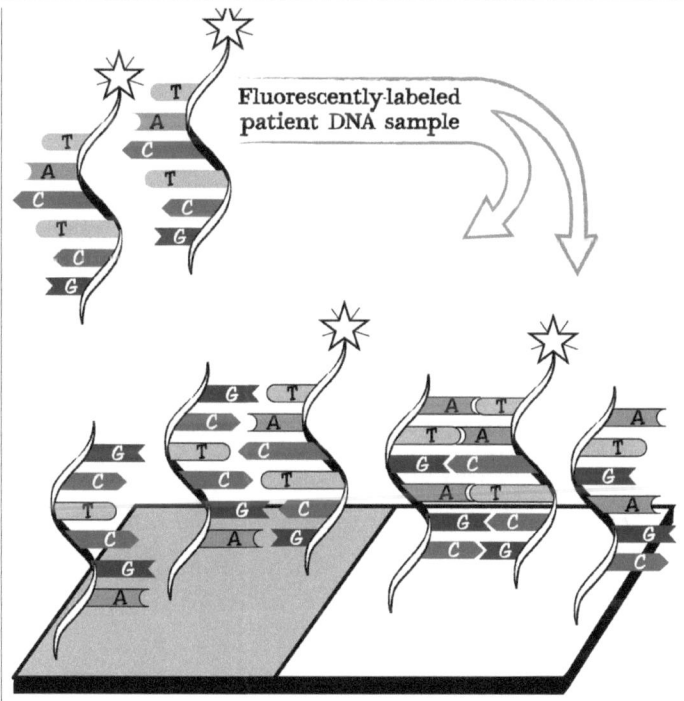

1. ISOLATE DNA: To begin, a technician collects a patient DNA sample, usually from a blood draw or cheek swab. The DNA is isolated. Samples generally get analyzed in the following order:

2. AMPLIFY: The technician copies the specific genes or DNA sequences under investigation, using the polymerase chain reaction (PCR). For example, if a doctor wants to see which **allele** or variant of apolipoprotein E (ApoE)—an Alzheimer's associated gene—a patient has, the lab technician amplifies that gene.

3. SORT and LABEL: The patient's DNA is then separated into single strands. The strands are marked with a fluorescent label the chip reader can recognize. The labeled strands are placed in a buffer solution for application to the chip.

4. LINE UP: Technicians then wash the labeled DNA onto the chip. If its bases complement any probes already there, the sample DNA will hybridize with those particular squares.

5. CLEAN and RECORD: Next, the technician washes off any DNA that didn't stick to the chip. The chip reader will detect any of the patient's DNA that remains.

6. SOFTWARE SCAN: Finally, researchers use a chip reading machine to identify the locations of labeled genes. This allows them to correlate the patient's DNA to a specific gene variant. Computer software keeps tabs on DNA positions and corresponding genes.

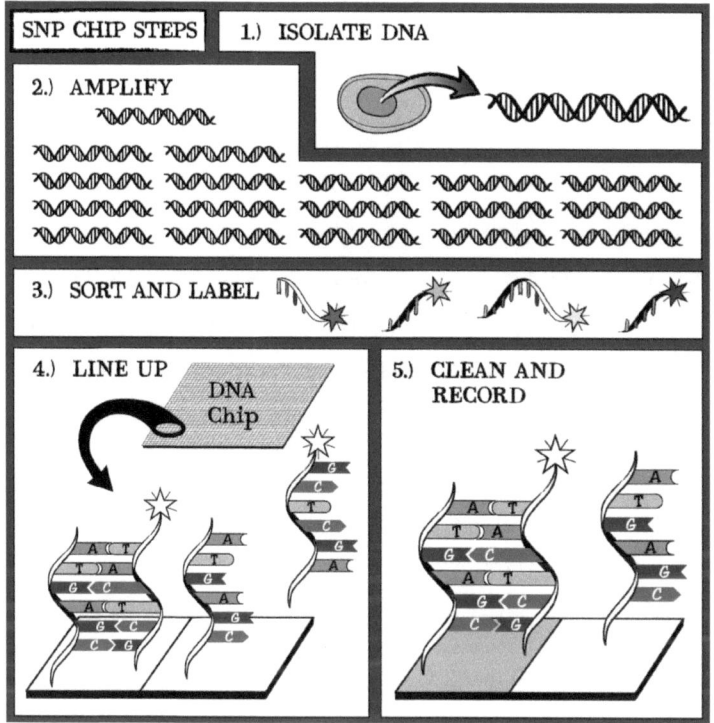

Chips at Work: Direct-to-Consumer Genetic Tests

SNP chips provide a relatively quick and inexpensive method to discover whether someone carries SNP variants associated with a given illness, such as atherosclerosis or celiac disease. This kind of test is sometimes called "direct-to-consumer (DTC)" genetic testing. That's because several companies, including 23andMe and Helix, offer this product to the general public. Consumers provide a DNA sample (usually by spitting into a prepared tube) and mail it to the company for processing. Results come via email. These tests *aren't* diagnostic. Instead, they look at *susceptibility genes*. They simply assess someone's risk of developing a certain disease as increased, decreased, or average.

For example, the ApoE gene is a particularly well-characterized susceptibility gene. One allele, ApoE4, is associated with a *higher risk* of Alzheimer's disease (AD). People with two copies of ApoE4 have as much as *twenty times* the risk of developing AD. However, some individuals with two copies *never* develop Alzheimer's according to statistical studies of different ApoE variants. On the other hand, another version of the gene—ApoE2—may reduce the risk of developing AD.

By identifying a person's ApoE gene variant, researchers can categorize his chances of developing AD—again increased, decreased or average risk. However, this test doesn't definitively diagnose or predict the disease's onset.

Clinical Studies

SNP chips are also used in clinical studies. For example, DNA from people with identifiable, disease-associated symptoms—say those associated with irritable bowel disease (IBD)—may allow researchers to better understand the illness. For example, a SNP chip that shows different alleles of the same gene could help reveal particular variants associated with a certain disease. Researchers compare DNA samples from groups of patients, such as IBD sufferers, to samples from healthy individuals. Then they investigate which variants occur more often in the IBD patients versus the healthy population and vice versa to look for patterns.

SNP testing is also routinely used in clinical studies of new treatments. Participants receiving the experimental drug also undergo a SNP analysis. This tests for SNPs that could affect the drug's safety and efficacy. These might include genes that code for proteins involved in drug metabolism or variations in the protein that the drug targets. If a participant experiences side effects during Phase I testing, researchers will analyze his SNP profiles. If patterns emerge, they can be used to include or exclude patients from clinical testing moving forward. The strategy also helps researchers examine a treatment's effectiveness in Phase II trials.

A Dash of Difference

Every human being alive shares an astounding 99.9 percent of their DNA. Only one tenth of one percent of genetic "stuff" accounts for a planetful of differences. That's because our genome consists of an astonishing three billion-plus building blocks—those As, Cs, Ts and Gs.

Areas of variation in genes are called **genetic markers**. Geneticists and others use them to assess the probability that people share a common ancestor. The more genetic markers they have in common, the more likely they're related. Only identical twins possess the same exact genetic markers.

Alphabet Soup: SNPs and STRSs

The two major types of genetic markers are **SNPs** and **STRs**. SNPs are single base changes in the DNA sequence. STR stands for "**short tandem repeat**." These small sequences of DNA repeat from five to fifty times. The number of times a particular STR repeats varies by individual.

This DNA is Not That DNA

SNPs and STRs give different types of information depending on the kind of DNA from which they originate.

- The **sex chromosomes** are the famous X and the Y from high school biology. Men have one Y chromosome and one X chromosome; women have two Xs. Thus, the Y chromosome contains DNA information about paternal ancestry.

- **Mitochondrial DNA (mtDNA)** is found in tiny compartments called mitochondria that convert sugar to energy in cells. Only women pass on mtDNA because the DNA in sperm mitochondria gets quickly destroyed during fertilization. Mitochondrial DNA can help trace maternal ancestry.

- **Autosomal DNA** is all the other genetic material. It's found in the pairs of autosomal chromosomes numbered one through 22. These pairs come from both mom and dad—one copy from each. Autosomal DNA provides information about both maternal and paternal lineages.

GOING FURTHER: DNA Ancestry

People used to rely on paper birth certificates, marriage licenses, and memory to discover their origins. Paper gets damaged, people are fallible, and memories fade. Leave it to biotech to come up with another way. New heritage-hunting techniques come courtesy of our increased knowledge of the human genome.

Deep Ancestry

Not only are different types of DNA inherited differently, but the rate at which their sequence changes from generation to generation varies too. That's because mtDNA and Y chromosomal DNA change only through random mutation of their As, Cs, Ts and Gs. Change can take centuries before a noticeable difference appears. By comparing genetic markers from mtDNA and Y chromosomal DNA with those of indigenous peoples around the globe, it's possible to estimate where your great-great-great great grandmother hailed from back in the day.

In contrast, autosomal DNA changes significantly with each generation due to **recombination,** which happens during the production of egg and sperm cells (germ cells). During this process, also known as **meiosis**, the copy of a chromosome that you got from your mom pairs up with the copy of the matching chromosome from your dad. This happens for each of the other 21 pairs. They then "swap" DNA content from corresponding positions, resulting in brand new chromosomes that contain DNA from each parental chromosome. So each egg or sperm cell contains completely unique chromosomes.

It's well known that siblings share half of their DNA with each other and half of their parents DNA. However, and this is the important bit—it's not the same half from sibling to sibling. That accounts for Jane having Mom's awesome nose and Joe getting Dad's "interesting" schnozzola. We share about a quarter with grandparents or grandchildren, aunts and uncles, nieces and nephews, and half-siblings. The amount of DNA we hold in common with relatives diminishes by half with each generation. By seven generations back, the amount of DNA shared among relatives amounts to less than one percent. In contrast to mtDNA and Y chromosomal DNA, autosomal DNA can only help make connections between living relatives or determine how far back you share an ancestor with someone.

Cocktail Fodder

Same Family, Different Ancestor?

Can siblings' ancestral origins differ? The surprising answer is yes! Imagine that 10 percent of a woman's DNA contains genetic markers for Mongolian ancestry. Because each egg carries only half a woman's DNA, only some would carry Mongolian markers. Therefore, one sibling could show Mongolian ancestry, while another doesn't.

Cocktail Fodder

Our Possibly Hairy Past

What's one of the most intriguing secrets genetic testing can reveal? Just how much of a Neanderthal you really are! Our smaller, sturdier cousins' genome was published in 2010, based on some very well-preserved DNA from bones found in a Croatian cave. The evidence suggests that early humans mated with Homo neanderthalensis. Thanks to DNA testing, we can find out just how cozy our distant relatives were with Neanderthals. The average person is about five percent Neanderthal.

GOING FURTHER: Proteomics and Other "Omics"

Understanding our genomes provides only one step in understanding health and disease. Molecular biologists have also begun characterizing the **proteome**, the full set of proteins expressed by a cell or organism. It's a big job. In fact, it's a much larger undertaking than sequencing the human genome. Consider: every different *cell type* in your body has a different proteome. To further complicate matters, our proteome changes depending on our age, general health, and even the time of day. The table below lists some other "omes" that have piqued researchers' curiosity:

OME	DEFINITION
Transcriptome	The entire set of RNA molecules produced by a genome, cell, tissue, or person at a given time.
Metabolome	The entire set of small-molecule metabolites produced by a cell, tissue, or organism at a certain time.
Autoantibodyome	All autoantibodies produced by a cell, tissue, or organism at a certain time.
Microbiome	The totality of microbes, their genomes, and environmental interactions in a particular environment
Exposome	The measure of all the exposures to potentially damaging substances of an individual in a lifetime

Advances in technology have led to a revolution in our understanding of the human genome and its role in disease. We are still in the early stages of collecting and analyzing genomic data, but it is expected to change the way we understand, treat, and prevent disease.

CHAPTER 6

Genetic Engineering

The DNA of every living thing consists of the same building blocks. Consequently, the DNA from different organisms is compatible. This astounding universality makes the entire field of **genetic engineering** possible. And genetic engineering enables scientists to introduce entirely novel genes into existing organisms with far-reaching results.

It sounds like a sci-fi dream: improving on nature by tweaking the stuff of life.

But how? Amazingly, researchers can join DNA pieces from different organisms to create **recombinant DNA**. Genetic engineers insert new genes into crops to imbue traits like resistance to pests or more controversially, pesticides. They place human genes into other organisms to produce human proteins. Researchers developed the first biotech therapeutic, human insulin, using recombinant DNA in bacteria.

Transgenic Organisms

A **genetically modified organism (GMO)** or **genetically enhanced organism (GEO)** contains genetic material that's been changed via genetic engineering. These techniques, generally known as recombinant DNA technology, use DNA from different sources and combine them into one molecule to create new genes. Scientists transfer the new and improved DNA into other organisms. **Transgenic organisms**, a subset of GMOs, have been given DNA from another species.

Tricky Terms

When a recombinant **plasmid** is transferred into bacterial cells, the cells are said to be **transformed**. When a recombinant plasmid is transferred into other cell types, the cells are said to be **transfected**.

Transgenic Organisms: *What Don't They Do?*

Though sometimes controversial, transgenic organisms are used to improve human health. There are three general categories of transgenic organisms: microbes, plants, and animals.

The first transgenic organisms were humble, ubiquitous, and ultimately incredibly useful *E. coli* bacteria. Stanley Cohen and Herbert Boyer **transformed** them with recombinant DNA. The technology enabled the pair to establish biotech pioneer Genentech in 1976. So the invention of transgenic bacteria, in a sense, marks the birth of biotech. Transgenic bacteria have been used to produce vital therapeutic human proteins including insulin and growth hormone ever since.

Scientists have found environmental applications of transgenic microbes, too. New and improved bacteria clean up oil spills, remove heavy metals such as mercury from water or soil, and produce fuel. So-called microdiesel has even been produced by genetically-engineered bacteria from bulk plant material.

Transgenic plants are now used to make therapeutic proteins and vaccines and to create improved crops. **Genetically modified (GM)** plants account for a significant percentage of food and textile crops, particularly corn, soybean, cotton, and canola. The US grows most of them, but Argentina, Brazil, Canada, China, and India also cultivate substantial amounts. Genetic modifications can increase resistance to pests, herbicides, pesticides, and viruses. They can also bolster resilience to harsh environmental conditions.

Bioengineering can even increase a crop's nutritional value—a process called **biofortification**. This innovation may prove especially beneficial in regions with limited sources of high-quality protein. Food crops such as rice and potatoes are being modified to yield higher levels of iron, zinc, pro-vitamin A, and essential amino acids.

In animals, genetic engineering has led to transgenic mice that are essential models for understanding disease. In addition, other transgenic animals have become steady, renewable sources of medically or economically valuable proteins. For example, biologic drugs are produced in goat or rabbit milk.

Animal scientists have also been endeavoring to increase the nutritive value of livestock. They're making genetic modifications that result in leaner beef cattle and in pigs whose meat contains omega-3 fatty acids. We'll get to the techniques for creating genetically modified animals at the end of this chapter.

BIOPHARMA INNOVATION: XENOTRANSPLANTATION

Xenotransplantation is the therapeutic use of animal organs in people. Its major drawback is the strong possibility of tissue rejection. A person's immune system recognizes the new organ as foreign and attacks it within minutes. Biopharma is introducing human "identity" genes into pigs (which have organs similar in size to ours) that cause the expression of human proteins on the outside of transplanted organs. If successful, the incidence of tissue rejection could decrease drastically, helping to alleviate the high demand for donated organs.

Bacterial Scissors?

In the early 1970s, scientists discovered a bacterial enzyme that could cut, like molecular scissors, through DNA strands at particular base sequences.

Called **restriction enzymes**, these powerful proteins enabled researchers to cut DNA from different species—for instance, from bacteria and human cells—and piece them together. For glue, they used another bacterial enzyme, DNA ligase. The hybrid DNA molecule was christened recombinant DNA, making biotech history.

RESTRICTION ENZYMES

Restriction enzymes cut DNA strands like molecular scissors along specific gene sequences.

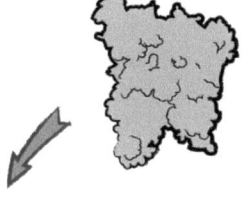

1.
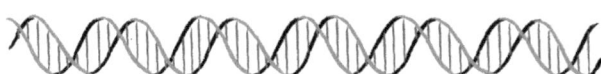

2.
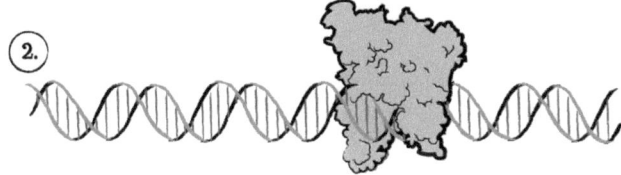

3.
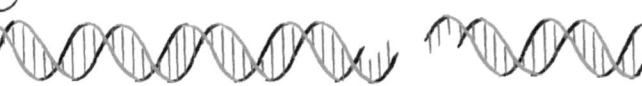

Each restriction enzyme recognizes a specific code sequence and then cuts within the site:

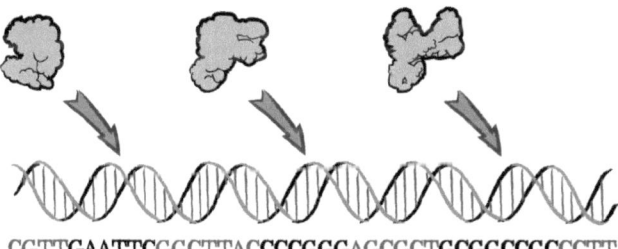

CGTT**GAATTC**GGCTTAC**CCCGGG**AGCGCT**GCGGCCGC**CCTT
CCAA**CTTAAG**CCGAATG**GGGCCC**TCGCGA**CGCCGGCG**GGAA

   EcoR1   Sma1   Not1
   sequence sequence sequence

 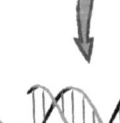 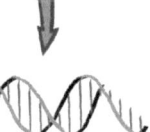 

CGTTG AATTCGGCTTACCCC GGGAGCGCTGC GGCCGCCCTT
CCAA**CTTAA** GCCGAATG**GGG** **CCC**TCGCGA**CGCCGG** **CG**GGAA

Chapter 6: Genetic Engineering

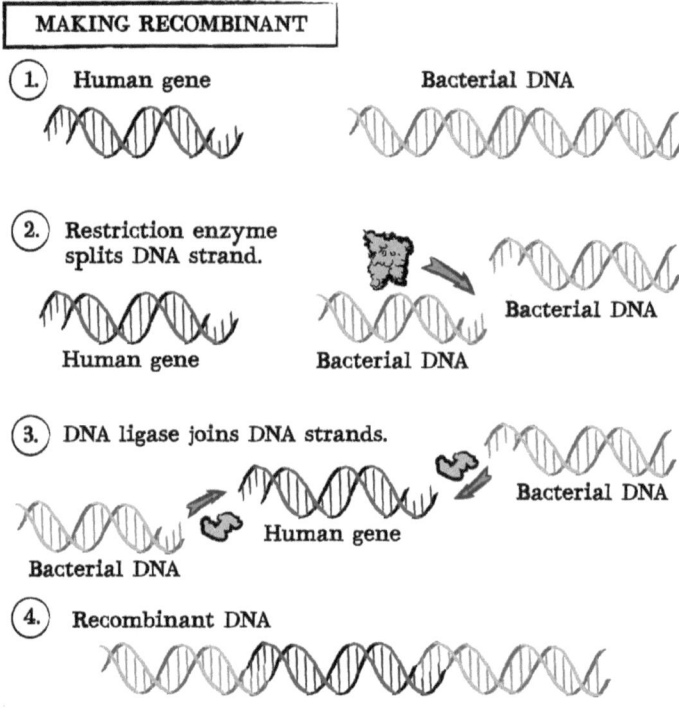

The bacterial DNA most often used for recombinant DNA is a **plasmid**. This small, circular DNA molecule can replicate independently from chromosomal DNA. Plasmids occur naturally in bacteria, where they increase genetic diversity because they can be transferred from microbe to microbe. Scientists use plasmids as **vectors** to carry **genes of interest** into bacteria.

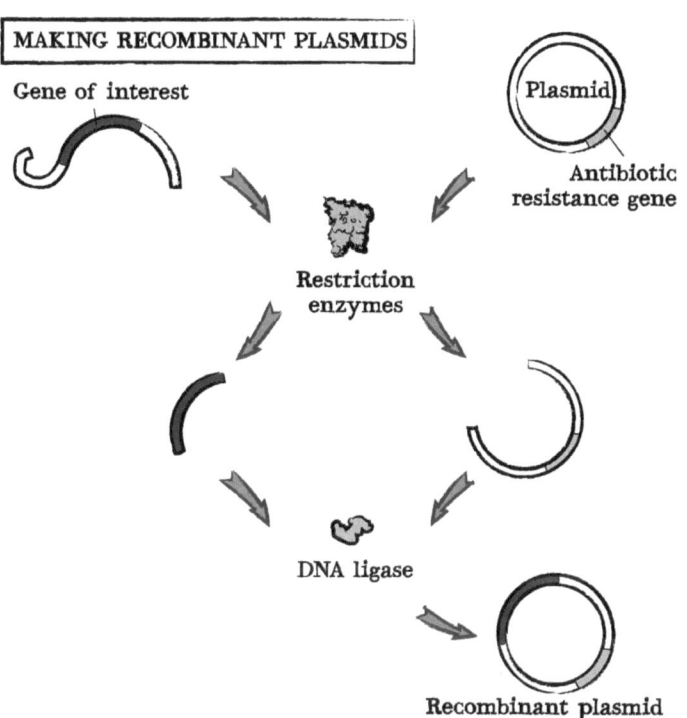

Tricky Terms

The term "recombinant" sounds highly technical. It really just describes the process of **combining** DNA from different sources to produce a new DNA molecule.

Bacteria Factories

Once a researcher creates a plasmid with a particular gene, he or she can integrate it into bacteria through **transformation**. Transformation is the genetic alteration of a cell resulting from the uptake (genomic incorporation) and expression of foreign genetic material. It can be done in two ways: 1) with chemicals that change the permeability of the cell membrane or 2) by **electroporation**, which creates small, transient holes in the cell membrane after electrical shock.

Chapter 6: Genetic Engineering

Transformation is never 100% efficient—there's always a portion of the bacteria that fail to take up the plasmid. Some species divide as often as every 20 minutes, so it's important to quickly establish a culture with bacteria containing only the introduced plasmid. Otherwise, the recombinant bacteria may soon be overwhelmed by non-plasmid containing bacteria.

Scientists exclude undesirable bacteria by including an antibiotic resistance gene in the plasmid. It produces a protein to make the bacterium resistant to antibiotics such as ampicillin, neomycin, or tetracycline. Researchers then raise the transformed bacteria in **growth medium** that contains the antibiotic. That way, only the right bacteria survive. Then they grow and divide, spawning many duplicate daughter cells with additional genetic information from the plasmid.

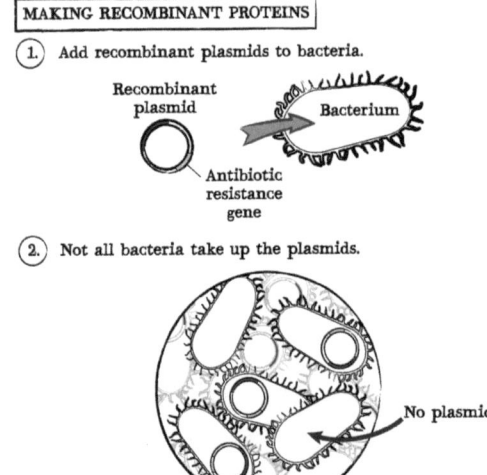

MAKING RECOMBINANT PROTEINS
1. Add recombinant plasmids to bacteria.
2. Not all bacteria take up the plasmids.

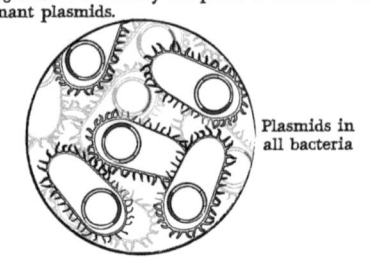

3. Add antibiotic to kill off bacteria lacking antibiotic resistance gene.
4. Surviving culture uniformly composed of bacteria with recombinant plasmids.

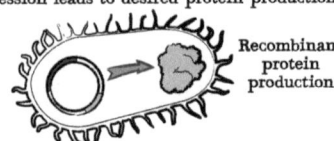

5. Gene expression leads to desired protein production.

Once the transformed bacteria start growing, they make recombinant protein from the plasmid gene, just as they manufactured the protein that confers antibiotic resistance.

BIOPHARMA INNOVATION: Manufacturing Insulin

A glance at current insulin production sheds light on the manufacture of recombinant proteins. Scientists insert the human insulin gene into a tetracycline-resistant expression plasmid and then introduce it into *E. coli*. The transformed *E. coli* grow on petri dishes in agar containing tetracycline to kill the non-plasmid bacteria. They then go into flasks containing nutrient broth that are kept at a comfy 37 degrees Celsius and shaken to ensure oxygen circulation. Next, they're transferred to fermentation vessels for large scale culture. The vessels are sealed, and the bacteria nurtured at a carefully controlled temperature with internal stirring. The bacteria are collected when they reach the desired density. Their cell walls are dissolved with enzymes and chemicals. The bacteria extract is passed through **chromatography** columns to purify the insulin protein. Column chromatography is widely used to separate proteins based on chemical and physical differences. (More on that in Chapter 9.) At the end of chromatography, scientists recover the pure human insulin.

Engineering Mammalian Cells

Cocktail Fodder

Dolly the sheep (1996-2003, RIP) was the best-known cloned animal because she was the first *mammal* to be cloned. But she was not the first *animal.* Way back in the 1950s, a tadpole was the first successful clone.

Bacterias are very useful for **molecular cloning**, making many copies of a particular gene. They are also useful for **expression cloning**, producing multiple copies of small proteins. Bacteria are essentially teeny tiny factories that create DNA and proteins. However, bacteria cannot help scientists understand human biology or human disease; that requires mammalian cells-human or otherwise. How do we "humanize" mammalian cells? Good ole' genetic engineering to the rescue.

Genetically engineered mammalian cell lines are versatile. They can be powerful models to study the control of gene expression, normal cell function, and the molecular and cellular mechanisms of disease.

Scientists also use genetically engineered mammalian cell lines to produce protein therapeutics. Mammalian cells offer some advantages over bacteria in protein production. Human proteins produced by mammalian cells are often more soluble, more properly folded, and more structurally sound than those from bacteria cells. Those qualities matter when it comes to manufacturing human recombinant proteins for use as therapeutics.

Scientists grow and maintain mammalian cells through cell or **tissue culture**. Scientists collect cells to culture in various ways. They can come from patients—think biopsy or even just a skin scraping. They can also come from mice or other experimental animals. The sample is treated with **trypsin,** an enzyme that breaks down the connections between cells to release individual cells. Researchers culture cells in media that mimic their typical environment, including nutrients and growth factors. The cells grow at 37 degrees Celsius, normal human body temperature. Unfortunately, the conditions that favor experimental cell growth also favor that of contaminating bacteria, yeast, and molds. That means all work with tissue culture has to take place in sterile laminar flow hoods. The hoods blow air away from the work area to diminish contamination. Even under favorable conditions, most cells divide only a few times before they die. Cancer cells, always the exception, grow well in culture. They've mutated to avoid cell death, and many cultured cells originally come from tumors. Once cells survive in culture for a certain period, about six months, they're considered a **cell line**.

Cell lines can be frozen and stored in liquid nitrogen. Researchers thaw them for experiments, although some die along the way. The survivors start growing and replicating again once they thaw, if they're well taken care of. In practice, most researchers obtain cell lines that are already established and characterized, from suppliers such as the American Type Culture Collection.

INDUSTRY NOTE: Storing Cell Lines

Large companies often store banks of frozen cells in different geographic locations as a precaution against losing valuable cell lines. One cell bank might be kept in Boston, one in San Diego, and a third in Germany.

Researchers insert genes into cultured mammalian cells with plasmids in a manner similar to bacterial transformation. This process, however, is called **transfection**. Scientists often use lipid reagents that enable DNA to cross the cell membrane around the cell. This way of introducing DNA is "gentler" than that used for bacterial cell transformation. The extra TLC is necessary because mammalian cells are more fragile than bacterial. The method includes transfecting the DNA into **liposomes**, small, membrane-bounded bodies that are in some ways similar to the cells themselves and can fuse with the cell membrane, releasing the DNA inside of the mammalian cell.

INDUSTRY NOTE: Mammalian Cell Lines

Mammalian doesn't necessarily mean human. Plenty of furrier mammalian cell lines make suitable research and manufacturing tools. Those most commonly used to produce recombinant proteins are Chinese hamster ovary (CHO) cells and mouse myeloma (NS0) cells.

Recombinant Proteins in Healthcare

As therapeutic agents, recombinant proteins are extremely versatile. Some are recombinant human proteins that replace or supplement those depleted or lost completely in the body. Examples include recombinant insulin for Type 1 diabetes, growth hormone for growth hormone deficiency, and interferon beta-1 to modulate the function of immune cells in multiple sclerosis (MS).

There's also erythropoietin or Epogen. This Amgen protein stimulates blood cells to proliferate during chemotherapy or kidney disease. Even some vaccines are now simply viral proteins made by using recombinant DNA technology.

BIOPHARMA INNOVATION: Gardasil

Gardasil, the human papillomavirus (HPV) vaccine, is produced by making a recombinant HPV protein called L1. Its molecules assemble to make virus-like particles or VLPs. The particles mimic the real deal thoroughly enough to spur the immune system into making HPV antibodies.

Tricky Terms

Many of today's antibody therapeutics are **humanized antibodies.** Scientists engineer a mouse antibody gene so that most of the antibody molecule is human and only a small part, called the variable region, is mouse. Humanized antibodies minimize the chances a therapeutic is rejected by the immune system.

Animal Farm

Domestic animals that produce milk, such as cows, goats, and sheep do double therapeutic duty. They can be induced to make human proteins and make them *only* in their milk, so these proteins are easily renewable! Also, because the milk-secreting system is separate from the rest of her body, she can make proteins like insulin or blood clotting factors which would otherwise kill her if released into her bloodstream.

Getting the Goat

To build a transgenic goat, scientists first isolate and clone the human gene of interest. Then they **ligate** (glue) with the enzyme ligase copies of the gene to a goat **promoter** sequence. A promoter is a stretch of DNA that controls whether or not a gene is activated in a specific cell type. By choosing an udder-specific promoter, scientists ensure that the gene will be turned on only in the transgenic goat's udder. Scientists then **microinject** the recombinant DNA into the nucleus of a fertilized goat egg. The egg is transplanted into a surrogate, who gives birth to a transgenic goat that expresses the target gene in her milk. The milk is collected, and the human protein isolated. More on just how to come in a later chapter.

Chapter 6: Genetic Engineering

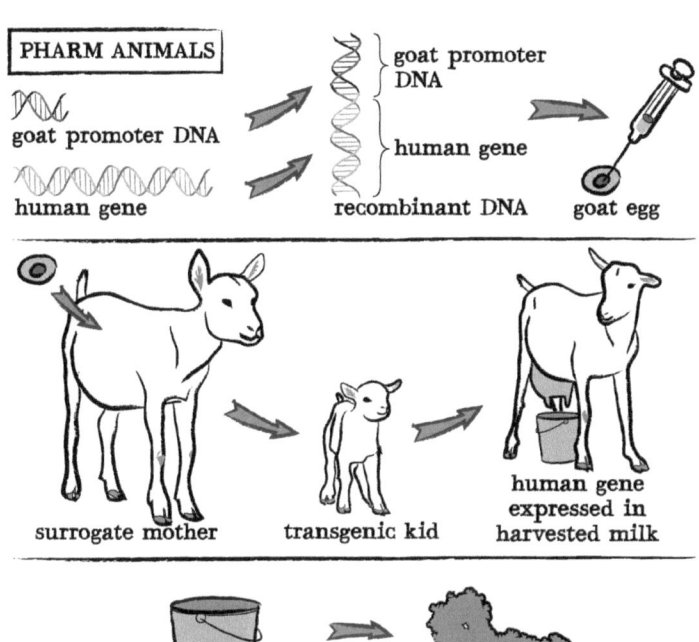

Cocktail Fodder

According to researches, DNA has a half-life of 521 years. This simply means that the oldest animal or organism that can be cloned back to life cannot be older than 2 million years. Thus, replicating dinosaurs is literally impossible because they became extinct 65 million years ago.

BIOPHARMA INNOVATION: Transgenic Animals

ATryn was the first FDA-approved biologic drug to be produced using transgenic animals. It's a recombinant version of an anti-coagulation factor, antithrombin. Ultra-modern medicine made possible by goats.

Magic Mice

Transgenic technology, mostly in mice, generates crucial models for drug discovery and development. By introducing our disease genes into mice, scientists create simulations to use for developing and testing drugs. Researchers also use the transgenic rodents to observe how genes operate in development, physiology, and disease.

BIOPHARMA INNOVATION: Fusions Proteins

Fusion proteins are recombinant proteins created by genetic engineering techniques that meld desirable properties from two different proteins. For example, hemophiliacs now receive long-acting recombinant blood clotting factors to improve their quality of life. These consist of the clotting factor fused with an antibody's **Fc region**. The Fc region gives antibodies unusually high stability, which enables them to remain in circulation up to several weeks. Fusing the Fc region with clotting factors buys hemophiliacs a few days of stability—long enough to reduce the treatment frequency to once every four days, versus every other day. Given that most receive factor VIII treatment intravenously, this advancement promises to make life better for hemophiliacs who need treatment.

In this chapter, we've described how scientists engineer cells to produce proteins. Next, we'll describe the human immune response and ways in which biotech companies apply an understanding of this response to their efforts.

CHAPTER 7
Drug Discovery

For millennia, people have sought natural ways to alleviate aches, pains, and illness. In the 3rd century BC, the ancient Greek physician Hippocrates described how chewing on willow bark could alleviate pain. In the late 19th century Bayer modified salicylic acid, found in the bark of the willow plant, to develop the drug Aspirin. Quinine, a component of the cinchona (quina-quina) tree bark, was used to treat malaria from as early as the 1600s by the native Peruvians. In Europe during the 19th century the bark was dried, ground, and mixed into wine before being drunk. In 1820, quinine was extracted from the bark and isolated. This purified quinine then replaced bark as the standard treatment for malaria.

By the mid-20th century drug development became less serendipitous, and scientists were developing drugs by trial and error. For instance, to find an aspirin alternative, British scientists at the Boot Laboratories tested hundreds of unrelated chemicals in guinea pigs to look for drugs that

Cocktail Fodder

Digitalis, or digoxin, which treats heart conditions, comes from the foxglove plant.

would reduce inflammation. This kind of research took lots of time and money with erratic success.

Beginning in the 1980s, scientists started to see drug discovery differently, adopting **rational drug discovery** or **mechanism-based drug design**. Researchers seek first to understand a disease at the cellular level, and then identify its mechanism. Most diseases can be traced to the underproduction of a protein (e.g., insulin in Type 1 diabetes), the overproduction of protein (e.g., the growth factor receptor HER2 in certain types of breast cancer), or the production of a mutated version of a protein that malfunctions. Understanding what protein is associated with a disease helps researchers design drugs to target or replace those proteins. Although still labor and resource intensive, this approach succeeds more often in producing highly effective, specific drug therapies for a given indication than the past methods of trial and error.

BIOPHARMA INNOVATION: Fighting HIV

Further refinement in rational drug design has incorporated structural information about the target protein. This change led to the invention of HIV protease inhibitors. The HIV protease is a viral protein. It cleaves other HIV proteins prior to viral assembly inside a host and is critical to HIV's survival. With this understanding, researchers realized they could stop the virus from spreading by inhibiting the protease's activity. This knowledge led in part to today's highly effective inhibitors.

In the 1990s, chemists developed huge collections (also known as libraries) of related chemical compounds using **combinatorial chemistry**. This approach helps researchers develop new molecules by synthesizing every possible chemical derivative of an existing molecule. It makes for very, very large libraries. A related approach, **directed evolution**, has generated new biologics. Researchers introduce specific mutations to a gene sequence to create new, but structurally related, proteins.

These chemical compound and biologic libraries make possible **high throughput screening** to find optimal drugs. For example, researchers might examine a vast chemical library for the molecules that best inhibit the virus HIV. In some sense, this is an integrated approach to drug discovery. It combines the trial-and-error approach of screening innumerable compounds with a mechanism-based approach of designing screening **assays** based on how diseases operate. (An assay is the scientific term for test.) In drug discovery, assays measure the potential efficacy of a drug candidate.

Chapter 7: Drug Discovery

Tricky Terms

<u>High Throughput Screening</u> is a method of drug discovery that uses robotics, data processing/control software, liquid handling devices, and sensitive detectors to quickly conduct millions of chemical, genetic, or pharmacological tests to identify active compounds, antibodies, or genes that affect a particular disease-associated biomolecular pathway.

Serendipity also plays a role in medicine. When Alexander Fleming returned from vacation to find his petri dishes of bacteria contaminated with mold, he didn't toss them. Instead, he noticed that mold seemed to prevent bacterial growth and surmised that the *Penicillium* mold must be producing *something* to kill bacteria. Eureka! Fleming had discovered the first antibiotic, penicillin, by accident.

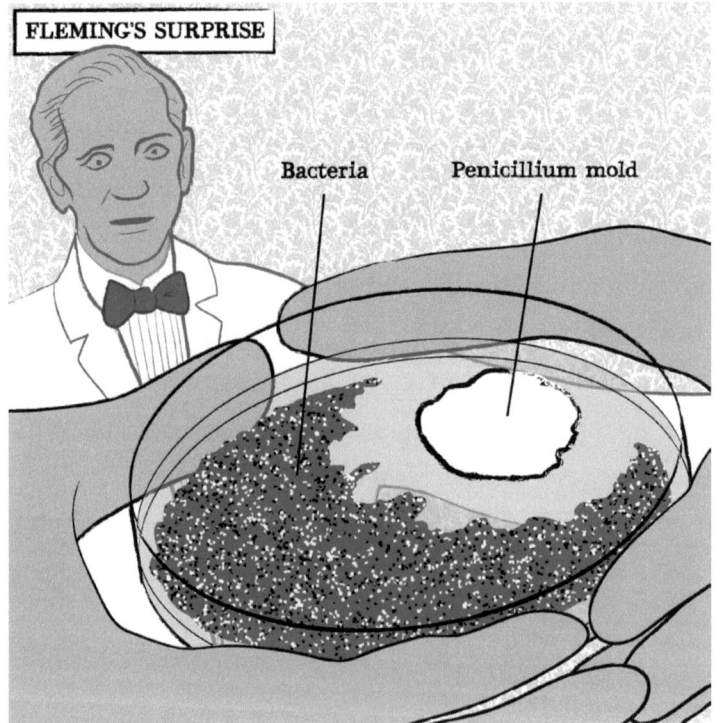

FLEMING'S SURPRISE — Bacteria, Penicillium mold

Identifying the Target

The first and arguably most important step in successful drug discovery is unraveling the molecular basis of the disease—determining the **drug target**. This step relies on years of basic research of a disease's biology.

Look at coronary artery disease (CAD). The most serious and life-threatening cardiovascular diseases are caused by atherosclerosis. This build-up of cholesterol plaque in arteries restricts blood flow and can lead to heart attack and stroke. Atherosclerosis can be caused by low-density lipoprotein (LDL), known as the bad cholesterol.

Although lots of Americans consume foods high in cholesterol, diet isn't its only source. Cholesterol is also synthesized in the liver. While many people can control it through diet and exercise, others cope with overactive cholesterol synthesis enzymes. That means they're prone to high cholesterol levels regardless of what they eat. Identifying the key enzyme required for cholesterol synthesis, HMG-CoA reductase, allowed scientists to develop drugs to block cholesterol synthesis.

Cocktail Fodder

Pfizer scientists were working on a new high blood pressure drug, Sildenafil, and brought it into clinical trials. It was a fail, at least for hypertension. *However,* many male subjects noted a dramatic side effect. Pfizer renamed the product Viagra, the first drug for erectile dysfunction. It's been a huge success.

BIOPHARMA INNOVATION: Statins

Collectively, drugs that inhibit HMG-CoA reductase are called **statins**. They include Crestor, Zocor, and Lipitor.

Tricky Terms

A drug target is the organ, tissue, or molecule involved in a disease that is modified or affected by a potential therapeutic.

Types of Targets

James Black, winner of 1988 Nobel Prize in Physiology or Medicine for discovering the beta blocker propranolol, has commented, "The most fruitful basis for the discovery of a new drug is to start with an old drug."

Black knew that some classes of drug target promise greater success than others. If a drug against one target works well, it's likely that a similar drug against a similar target will also be effective.

Among target classes, G protein-coupled receptors (GPCRs) have been the most successful. Recall from Chapter 2 that G protein-coupled receptors are widely used in cells to regulate processes from blood pressure to nerve transmission to the secretion of stomach acid.

Researchers have also successfully developed drugs that target proteins and ion channels. In fact, more than half of all approved drugs focus on blocking GPCRs, receptors, or ion channels. For examples, the FDA has okayed dozens of kinase inhibitors, most as cancer treatments. Protein kinases, remember, play a critical role in growth factor signaling cascades.

BIOPHARMA INNOVATION: Inhibitors

GPCR-targeting drugs include:

- Beta blockers such as Atenolol inhibit epinephrine beta receptors to treat angina and hypertension.
- H2 inhibitors such as Zantac and Pepcid block histamine receptors to treat ulcers.
- H1 inhibitors such as Claritin and Allegra also shut down histamine receptors to treat allergy symptoms.
- Imitrex block serotonin receptors to alleviate migraines.
- Zyprexa plug up dopamine receptors which help schizophrenia and bipolar disorder manic episodes.

Validating the Target

When they identify a potential drug target, researchers try to validate it by determining its importance in the disease process. They also need to know whether targeting it is safe and effective. Target validation is critical to drug discovery. Research and development gets only more expensive as it proceeds. It doesn't take a Big Pharma CFO to see that it's ill-advised to pour millions of dollars into a dubious target.

Target validation most often includes cell-based assays (*in vitro* testing) and animal models (*in vivo* testing). Many therapeutic interventions seek to inhibit the activity of the selected target. Accordingly, many validation assays measure the effects of inhibition. Sometimes a target contributes to disease progression—but even if a drug inhibits

Tricky Terms

The terms *in vitro*, *ex vivo*, and *in vivo* indicate where an experiment takes place. All terms derive from Latin roots. *In vitro* translates to "within glass." It refers to studies in lab-grown cells. *Ex vivo* means "out of the living" and describes experiments on cells, tissues, or organs that have been taken (typically within 24 hours) from a living organism. *In vivo* translates to "within the living" and refers to live animal testing. A more contemporary (1989) term, *in silico*, is sometimes used to designate computer simulations or modeling.

it, another cellular protein may take over and cancel it out. In other cases, inhibiting a selected target may produce the desired outcome—halting cancer cell growth, for example—but produce unexpected side effects, such as the death of healthy cells.

One of the most popular ways to test the effects of inhibition is through **RNA inhibition (RNAi)**. Briefly, RNAi uses **short interfering RNA (siRNA)** to block the production of a protein. (Recall RNA contains the directions to make a protein. If there is no RNA, there is no protein.) It can quickly determine the results of blocking protein production, mimicking the effects of a strong inhibitor.

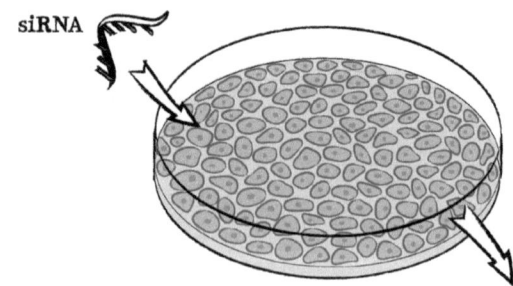

TARGET VALIDATION: RNAi

In vitro: cell models

1. Does the target play a key role in the disease process?
2. Is targeting it likely to be effective and safe?

- Will cancer cells die?
- Will nerve cells stay alive?
- Will beta cells make more insulin?
- Will liver cells make less cholesterol?

If cell models show promise, researchers move to animal models. They often experiment with "knockout" mice, in which a particular gene has been disrupted. The little furry knockouts enable researchers to ask questions similar to those posed in cell models, but with a whole animal. Do the mice still get cancer, Parkinson's disease, diabetes, or heart disease when the target gene is blocked or absent? The creatures also shed light on targeting safety that cell models might not address, because the effects of silencing a gene show up on a whole being.

GOING FURTHER: 3D Tissue Arrays

Currently, drug development necessitates *in vivo* testing to best understand how drugs act in a human body. *In vitro* testing just can't replicate Homo sapiens' complex physiology. However, a new innovation is helping to reduce the need for animal testing: 3D tissue arrays. Companies use 3D printing to create tissue arrays that better mimic human physiology than flat layers of cells in tissue culture flasks.

Therapeutic Choices: Small or Large?

The next step in drug discovery is designing a targeting strategy. This usually comes down to choosing between a small molecule drug or a large molecule drug also known as a biologic.

Small molecule drugs can usually cross cell membranes and enter cells; they target proteins *inside* cells. Small molecule drugs are also able to cross the blood-brain barrier. Their target specificity can vary—it may be good or even

Chapter 7: Drug Discovery

Tricky Terms

The **half-life** of a drug refers to its persistence in the blood or target tissues. The name refers to the fact that it measures how long it takes for the drug concentration to drop to one half of its initial levels.

very good. In some cases, though, the fit is only fair. That's a problem which could result in off-target toxicity. Finally, small molecules can be taken orally. Once someone ingests them, they generally have a short **half-life**. That means the liver's drug-metabolizing enzymes quickly break them down.

Biologics, in contrast, usually can't make it past cell membranes or the blood-brain barrier because of their large size. This limits them to targets *on the surface of* or *outside* of cells as well as outside the brain. Biologics generally have extremely high specificity, so they're less prone to interfere with proteins other than the target. Ideally, this decreases the risk of off-target toxicity. Because biologic drugs are proteins, they have to be injected or infused, not swallowed, because digestive enzymes break them down. Once inside the body, however, they're stable because the liver doesn't break them down.

BIOPHARMA INNOVATION: One Target Multiple Drugs

A few companies hope to change how patients take biologics. Check out these innovations:

- A "robotic pill" that protects biologics from stomach acids. The itty bitty robot "injects" the drug into the intestinal wall, delivering it right to the bloodstream.
- Delivering a gene for the biologic to intestinal cells, enabling *the body's own cells* to produce the drug. This approach is especially appealing for diseases affecting the colon and small intestine. Take that colon cancer or Crohn's disease!

Chapter 7: Drug Discovery

BIOPHARMA INNOVATION: Drug Choices

Some diseases can be treated with both small molecule drugs and biologics. This is particularly true for growth factor receptors, since part of the receptor is outside of the cell (extracellular) and part is inside of the cell (intracellular). Physicians can target the HER2 growth factor receptor with a biologic such as a monoclonal antibody, think Herceptin, as well as with small molecule inhibitor drugs such as Tykerb. Similarly, the epidermal growth factor (EGFR) receptors can be targeted by anti-EGFR antibodies such as Vectibix, as well as small molecule inhibitors such as Iressa.

Assay Development

To screen drugs, scientists design **assays** to identify the drug candidates with the most potential. A test must be fast, accurate, and amenable to scale-up. "Scale-upability" means that thousands of compounds can be screened efficiently. Often researchers develop assays that induce a fluorescent signal or other color change that are used to easily quantify results. Fluorescence is easily measured, relatively inexpensive, and safe. It also works with many different assay designs.

Glowing Results

Suppose scientists have identified a cellular enzyme as a drug target. Imagine that this enzyme is too active, and as a result, cells are dividing when they shouldn't, leading to cancer. Drug discovery scientists are trying to identify small molecules that will inhibit this enzyme using a fluorescent assay as described above. How exactly would such an assay work?

Cellular enzymes have active sites, pockets where chemical reactions are catalyzed. Active sites have specific structures that bind only to the enzyme's substrate, the molecule with which it interacts. Researchers can modify substrates to emit a fluorescent signal if the enzyme acts on them. This gives researchers a quick readout. Researchers would then add different potential small molecule inhibitors of the enzyme before adding the modified substrate. Enzymes treated with effective inhibitors (potential drugs) won't successfully catalyze the reaction so the substrate won't fluoresce. In other words, no fluorescence means the small molecule being tested successfully inhibited the overactive enzyme, and should be studied further to see if it can be developed into a new drug.

Chapter 7: Drug Discovery

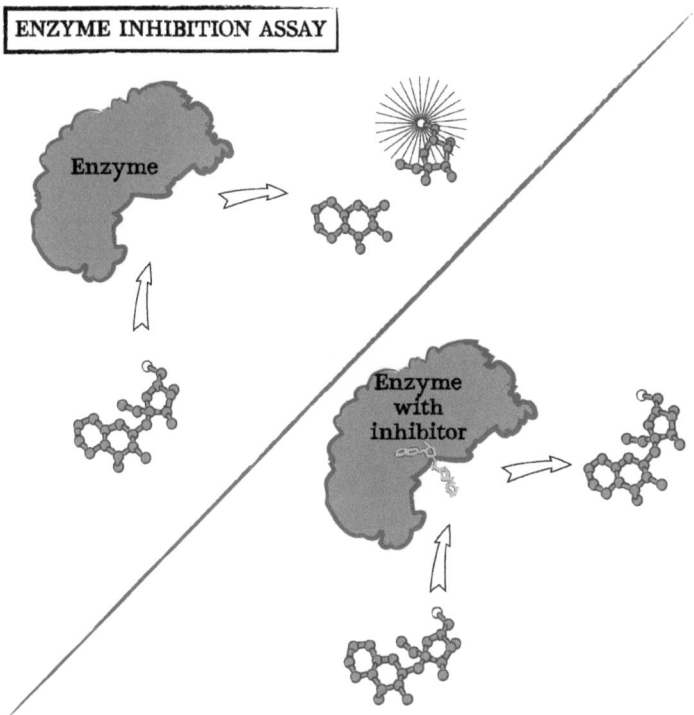

ENZYME INHIBITION ASSAY

Researchers can do nearly the same thing with a receptor. Scientists engineer cells to produce a fluorescent signal when a particular receptor activates. For example, a cell could be engineered to glow green when the breast cancer-associated HER2 receptor kicks in. Researchers would then add a potential HER2 inhibitor, followed by an appropriate HER2-activating growth factor, and measure fluorescence. If fluorescence appears, that means the receptor was activated, and the inhibitor was not effective. No fluorescence means that the potential inhibitor successfully prevented receptor activation, and should be explored further as a potential new drug.

Chapter 7: Drug Discovery

GOING FURTHER: Combinatorial Libraries

To select the best possible enzyme or receptor inhibitor, researchers screen as many potential candidate compounds as possible. Drug companies use large libraries of chemical compounds for high throughput screening assays. They're are often proprietary, constructed by the company explicitly for their own drug discovery programs.

There are different kinds of chemical libraries. **Combinatorial libraries** are assembled like pyramids. First, a small but diverse set of twenty chemical compounds are attached to small beads. These are "scaffold" molecules on which the library will be based. Each set of beads is split into fifty batches, which are reacted with a different chemical reagent to produce a slightly different compound. This generates one thousand unique first round compounds attached to beads. Next, these beads are pooled together and split into fifty batches, then reacted with a different set of fifty second-round reagents. The effort results in a library of 50,000 unique second-round compounds. Researchers can repeat the process until they generate millions of different compounds.

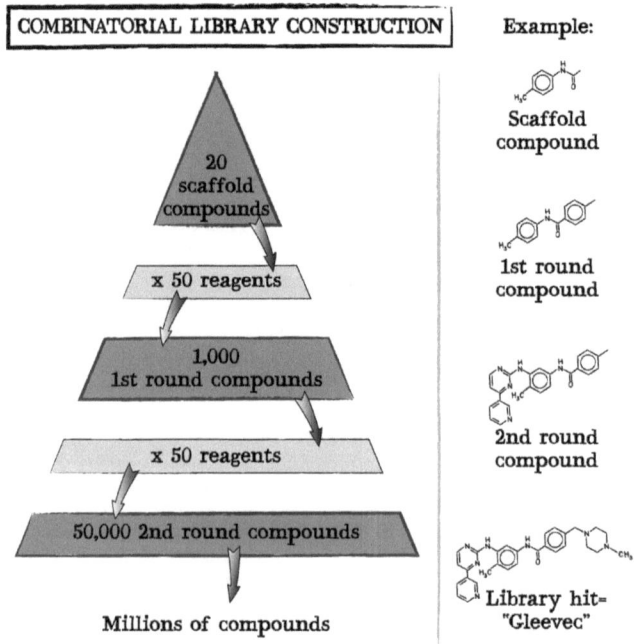

High Throughput Screening

With so much to test, scientists need to maximize efficiency. When developing assays, they rely on high-throughput screening. This usually take place in a micro-titer plate with 96 or 384 wells.

Let's say we know Enzyme X is causing a disease so we want to screen for an inhibitor of Enzyme X. To start, researchers add the Enzyme X to each well. Next, they add a different inhibitor to each well, one somewhat similar to the Enzyme X natural substrate. The hope is that one (or more) inhibitors fit into and plug up the Enzyme X active site. This plug will prevent binding with other ligands. To determine if the inhibitor binds to Enzyme X, a fluorescently-labeled substrate (as described above) goes into each well. Wells where the enzyme is still active will fluoresce. Any in which the enzyme is inhibited won't glow. Any "blank"—non-glowing—wells represent potential inhibitors. These potentially promising inhibitors are called "hits" and merit further investigation as a potential therapeutic.

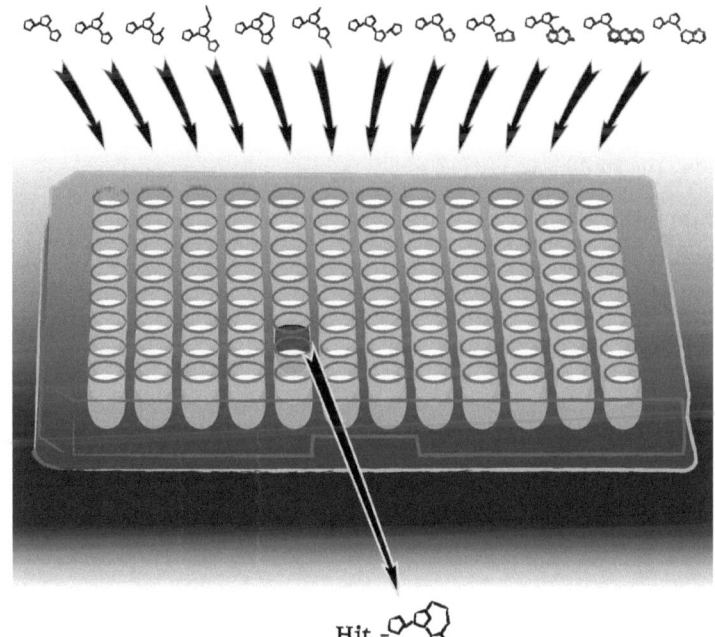

White indicates a fluorescent glow; dark indicates no glow. In this example "no glow" indicates a hit.

Combinatorial Library Success

Gleevec, which treats chronic myelogenous leukemia (CML), is the poster drug of rationally-designed anti-cancer treatments. The illness is caused by a rearrangement, or translocation, of two chromosomes in lymphocytes (white blood cells). The chromosome shuffle induces two genes—Bcr and Abl—to form a new gene Bcr-Abl. This hybrid gene produces a protein—part Bcr and part Abl protein. The Abl protein is a kinase involved in controlling cell division and survival. In normal lymphocytes, Abl is usually turned "off" (doesn't work), waiting for the appropriate stimulus to turn "on." When lymphocytes are exposed to their growth factors, Abl is switched on briefly to send the growth signal (by phosphorylating target proteins) and then is switched off.

Abl's attachment to Bcr in CML causes it to stay "on" permanently. Therefore, it phosphorylates more target proteins, more often. The permanent "on" signal leads to uncontrolled lymphocyte growth and eventually leukemia.

Researchers combed a combinatorial library of compounds for one to fit Bcr-Abl's active site, potentially blocking its ability to phosphorylate target proteins. They found imatinib, Gleevec, with fantastic results. First, the drug blocked Bcr-Abl's kinase activity. It also stopped the growth of cancer cells in the lab *and* the growth of tumors in animals. Looking at the structure of a Bcr-Abl crystal, researchers saw that imatinib fit into the active site of the Abl part of the molecule. The precise fit prevented the drug from blocking most other kinases, resulting in fewer side effects than a drug that would work with less specificity.

Imatinib's success was unprecedented! More than 90% of patients remain leukemia-free for five years! In fact because Gleevec was so successful in clinical trials, doctors put the patients taking placebo or another drug on Gleevec during the clinical studies.

Gleevec blocks both Bcr-Abl and normal Abl but doesn't really impact normal lymphocyte function. This suggests that tumor cells depend far more on their mutated oncogenes than non-tumor cells depend on their corresponding normal genes. Following quickly on Gleevec's heels, researchers identified many other kinase-blocking drugs that received FDA approval for cancer treatment. Most of these target kinases are over expressed or mutated in cancer.

BIOPHARMA INNOVATION: Kinase Inhibitors

More than one hundred other kinase-targeted cancer drugs are in clinical trials. Many of them focus on the intracellular kinases that are mutated in cancer. Researchers consider kinases "druggable." They understand their structure and so can synthesize molecules to fit into their active sites.

Other protein molecules, such as the transcription factors, also play important roles in cancer but are often considered "undruggable." We just don't yet know enough about their structure or how to block their function.

From Hit to Lead

At the end of high-throughput screening, researchers will ideally have identified several compounds that show promise. Contenders are called "hits." To become "leads" and progress to animal testing, "hits" face rigorous *in vitro* testing. These questions need answers:

1. Is the hit safe?
2. Is the hit specific to its target?
3. Is the hit effective in treating the disease?

If a hit is supposed to inhibit an enzyme—say a protein kinase—it needs to affect only *that* one enzyme. That's because many kinases are essential to normal cell growth. If the hit isn't adequately specific, it might provoke severe side effects. Protein kinases, like other families of enzymes, share common sequence elements and structures. Some degree of cross-reactivity, therefore, is expected. However, researchers develop hits into drugs that inhibit the target enzyme at much lower concentrations than would affect other related enzymes. This provides a **therapeutic window**, the concentration range within which drugs are effective and safe. Researchers often perform specificity studies by testing drugs for inhibition with a large panel of related enzymes.

BIOPHARMA INNOVATION

Research support companies supply panels of protein kinases for bulk specificity testing.

The next consideration in early testing phases is potential effectiveness. Scientists examine a drug's **efficacy** in cells, starting with cells from human patients or animals designed to mimic critical features of the disease, before moving on to whole animal models. From cell-model efficacy experiments, researchers derive an **EC50 value**. That's the concentration at which half a drug's maximum desired effect is observed. For instance, if a drug should kill tumor cells, then researchers know its EC50 value when half of them die. Cancer cell models suit these studies because they grow so nicely in culture.

Cell models also allow scientists to begin exploring the question of drug safety. Many drugs bind to, and block, a potassium ion channel expressed in the heart called hERG. This interference causes heart muscle function to change, which can lead to life-threatening arrhythmias. Consequently, several drugs have been withdrawn from the market. To test for hERG blocking, researchers use cell lines engineered to express hERG. The experimental drug is added to the cells, and instrumentation that measure electric current determines if the hERG ion channel is being inappropriately activated.

Cocktail Fodder

The acronym hERG stands for "human Ether-a-go-go Related Gene." (Really!) The "ether-a-go-go" gene was first identified in fruit flies. Scientists observed that mutations caused fruit flies to shake their legs like go-go dancers. Needless to say, they named the gene in the 1960s, the peak of go-go dancing's heyday.

Chapter 7: Drug Discovery

INDUSTRY NOTE: Drugging the Undruggable?

Targets which have been considered "undruggable" because of their cellular location or unknown structure may turn out to be druggable after all. A new class of treatment, antisense drugs, block the expression of a specific gene. That reduces levels of the corresponding protein. Antisense drugs potentially offer an even more potent treatment approach than strong inhibitors. *The Biopharmaceutical Primer: An Insider's Guide to Advanced Therapies* explores these and other nucleic acid-based therapies.

Biomarkers

Biomarkers are specific, measurable physical traits that can determine or indicate the effects or progress of a health condition. They're critical in drug discovery and development. LDL cholesterol is a classic example. It's been clearly linked to heart disease and makes an ideal biomarker because 1) it circulates in the blood and so is easily obtainable; 2) it's an early indicator of heart disease—a patient can have elevated LDL decades before developing significant atherosclerosis; and 3) it changes as the disease progresses. A change in LDL level can serve as a clinical indicator that an experimental drug is working during clinical trials, in lieu of waiting potentially decades to see if a patient has a heart attack or not.

Biomarkers also facilitate running shorter, less expensive clinical trials. Statins have been sold since 1987. We now know they increase life span because patients have been on them for over 30 years. Initial clinical trials, however, simply showed a decrease in blood cholesterol.

In this chapter, we've gone over the major steps required for successful drug discovery. In the next, we explore how new drugs gain market approval.

CHAPTER 8

Drug Development: From the Lab to the Clinic

A successful drug development project starts by identifying potential new drugs in the lab, proceeds to testing those compounds in animals and humans, and ends with FDA approval. During drug discovery, researchers may screen thousands, or even millions, of compounds and select a handful for further development. *One* of these drugs might just make it to market.

Regulatory Agencies

Drug candidates overcome countless hurdles before arriving in a drugstore or hospital. Most concern safety and effectiveness. The arduous, expensive path to approval is regulated and overseen by governmental agencies. In the US, Congress passes laws to regulate drug approval. This includes the Federal Food, Drug and Cosmetic (FD&C) Act of 1938. Legislators wanted to ensure patient safety after the death of more than a hundred people from a poisonous sulfanilamide drug. Congress later amended FD&C to include guidelines for effectiveness. This is the most far-reaching piece of drug approval legislation in the United States.

Most countries have a regulatory equivalent of the FDA. In the European Union, the **European Medicines Agency (EMA)**, headquartered in Amsterdam, is a collective regulatory body for EU members so that drug makers can file for drug approval centrally. This centralization also helps ensure the process remains as free as possible from prejudices that stem from an individual state's domestic interests.

Regulatory Step 1: Preclinical Trials

In the US, the discovery phase of drug development runs from two years to more than a decade. The time it takes for a drug candidate to become eligible for testing in human patients depends on many factors. Among them: the complexity of the disease in question, the drug's chemical characteristics—simple properties like drug solubility in water can affect development timelines—and human and other resources a company invests.

Preclinical trials—safety testing in animals—is typically considered the last phase of drug discovery. Companies are usually required to evaluate a drug candidate in at least two different species. This includes recording toxicology, pharmacokinetic, and pharmacodynamic data. Preclinical trials are governed by FDA Good Laboratory Practices.

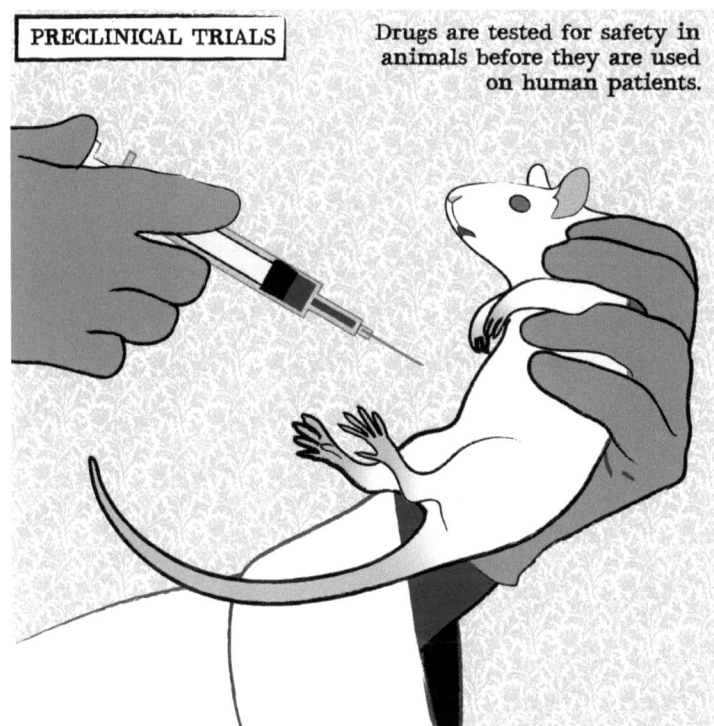

PRECLINICAL TRIALS

Drugs are tested for safety in animals before they are used on human patients.

Tricky Terms

Pharmacokinetics (PK) measures what the body does to a drug including absorption, distribution, breakdown, and excretion. Conversely, **pharmacodynamics (PD)** describes what the drug does to the body including how increasing concentrations of the drug influence potential toxicity. Both parameters are important in establishing safety and dosage.

After promising preclinical results, a company applies for **Investigational New Drug (IND)** status. Technically, the IND exempts manufacturers from the legal requirement that drugs be approved for marketing before shipping across state lines. In practice, IND status allows companies to begin clinical testing.

The FDA requires three main kinds of documentation for IND review: results from animal studies of drug pharmacology and toxicology, manufacturing information showing that the drug can be mass-produced consistently, and finally, detailed protocols describing the clinical trials. This last piece must specify the physicians who will oversee the trials and affirm that they'll follow informed consent procedures.

The FDA requires that clinical protocols clearly define endpoints because clinical trials measure endpoints or major health outcomes. There are generally two types. The gold standard are **clinical endpoints** which refer to benefits such as survival, decreased pain, the absence of disease, or greater mobility. The second type are **surrogate endpoints** which are substitutes for clinical endpoints when clinical endpoints are impossible or impractical to measure. The clinical benefit of survival, for example, should take decades to observe. Researchers may instead look to shorter-term phenomena. For example, in studying a drug for heart disease, clinicians can monitor cholesterol levels instead of decreased fatality from heart attacks. Similarly, in some cancer treatments, reduced tumor size stands in for longer life. Endpoints are front and center in the IND.

Regulatory Step 2: Clinical Trials

Trials of new drugs in humans evaluate efficacy—whether a drug cures or lessens the effects of disease—and safety. Each of the three phases of a clinical trial is governed by the FDA's standards for Good Clinical Practice.

Phase 1 tests drug safety in a small number of healthy volunteers. Under close supervision, patients receive escalating doses until side effects start to appear, at which point escalation stops. In this way, researchers establish a drug's **maximum tolerated dose (MTD)**, a benchmark for subsequent trials.

Some Phase I trials enroll patients instead of healthy volunteers. Many cancer drugs, for example, are deemed too potentially toxic to give volunteers. In addition, patients with cancer or another serious disease without available treatments may benefit from these initial trials. In either case, Phase I studies typically include between five and one hundred volunteers.

Phase II is where efficacy is first measured. Does the drug act as its developers intend in humans? Participants during this stage are patients only. Researchers assess drug efficacy by monitoring biomarkers of disease progression. In a cancer drug, this could be tumor growth. With a diabetes drug, it could be glucose tolerance. In this phase, researchers compare the drug's efficacy in the test group to a **control** group.

People in the control group receive one of two treatments: a "standard of care" treatment or a **placebo**. The standard of care drug is already approved and used for the same disease. The other option is a placebo, a treatment without actual medicine, such as a sugar pill. Researchers closely monitor for side effects throughout Phase II, and throughout the entire study. Phase II trials include more participants than the initial testing. Study size typically ranges between 50 and 1,000 volunteers. It can possibly be even more than 1,000, depending on the magnitude of the eventual patient population.

Tricky Terms

Researchers use two different statistical designs in Phase I trials. **Single ascending doses (sad)** design means that small groups of subjects receive a *single* dose of a drug. If researchers observe no adverse effects, a new group of subjects receives a higher dose, and so on until the maximum tolerable dose is determined.

Multiple Ascending Doses (MAD) design means the same group of subjects receive *multiple* low doses of the drug. Succeeding dose escalation for further groups is based on safety data.

Chapter 8: Drug Development: From the Lab to the Clinic

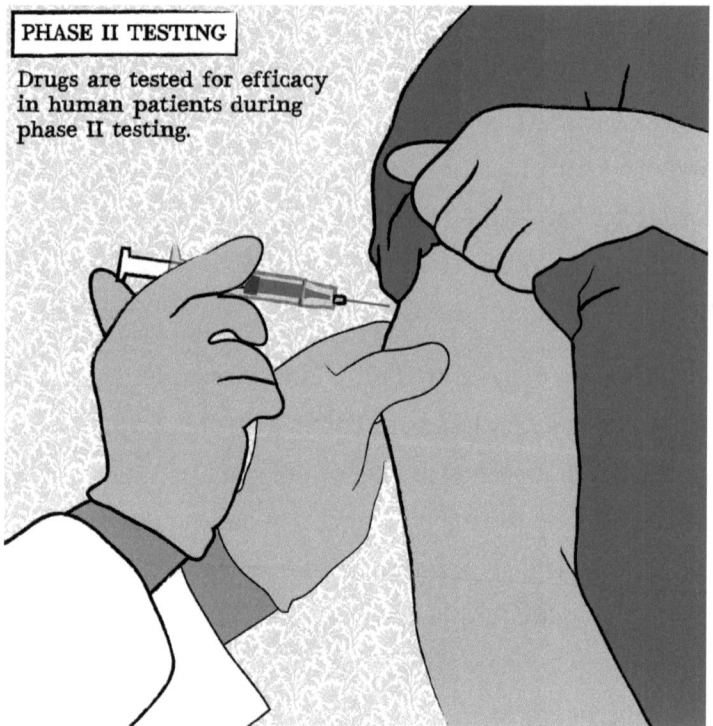

GOING FURTHER: Placebos

Scientists have found that placebos can significantly impact health. The well-known **placebo effect** may stem from an expectation or belief that a treatment has benefits. Because patients receiving the actual test drug usually expect the same (in addition to any effect the drug actually has), the placebo control is very important. Recent evidence shows that the bodies of people with chronic pain can produce **endogenous** opioid painkillers in response to a placebo.

More rarely, the reverse of the placebo effect occurs. Some patients anticipate that a drug won't help or may even harm them. When that happens, researchers call this the **nocebo effect**.

Phase III continues to test efficacy and safety, but with many more volunteers. A test group of 300 to 5,000 patients allows for more statistically significant analysis. Phase III trials (and Phase II) are usually **randomized, double-blind** studies. Patients are assigned randomly to one of three groups: drug, placebo, or standard of care. This randomization is one of the controls that helps eliminate bias that may occur in assigning groups.

In a double-blind experiment, neither participants nor researchers know who belongs in which group. A third party keeps a record of which subjects belong to each group, which doesn't get revealed to the researchers until the study ends. In some instances, it's impossible to perform a double-blind study—say in a study comparing surgical to non-surgical intervention. If a new drug seems effective during Phase III, patients are allowed to keep taking it after the study ends and before drug approval.

If the drug meets the expected clinical or surrogate endpoints during Phase III, its producers submit either a new drug application (NDA) for small molecule drugs or a large molecule drugs licensing application (BLA) for biologics to the FDA. Review times vary, but it typically takes about a year. There are three possible outcomes: the agency either grants approval, denies approval, or requests additional studies.

Even after approval, new drugs continue to be monitored. **Phase IV** seeks to catch any unexpected side effects of an approved drug. The FDA relies on patients, physicians, and the drug companies themselves to report

Tricky Terms

The World Health Organization defines **pharmacovigilance** as the science and activities relating to the detection, assessment, understanding and prevention of adverse effects or any other drug-related problem.

incidents. Unanticipated safety issues occasionally show up because the number of patients on an approved drug vastly outstrips the number of volunteers who tested it. If the side effects are serious enough, regulators or the company itself may pull the drug from the market.

GOING FURTHER: Real-World Evidence

Organized customer groups are now demanding evidence of value. This is increasingly achieved via **outcome-based reimbursement and real-world evidence**, in which patient health and outcomes data gathered outside of clinical trials drives downstream reimbursement.

New Kid In Town: Adaptive Design

Adaptive studies offer more flexibility than traditional designs, and are gaining in popularity because of their efficiency. They allow investigators to modify trial design as they go, rather than spending time and money pursuing drug formulations or dosages that ultimately prove ineffective. For example, researchers may separate participants into different dosage groups. At a prespecified time, they note patients' progress. If one dosage seems more effective than another, the trials continue with only that more effective dosage. Researchers set the trial protocol, which lays out the adaptation schedule and processes before the trial begins.

The 21st Century Cures Act promotes adaptive trial design. It benefits volunteers because researchers assign more patients to study groups taking medicine that shows promise. It's good news for non-study patients too. Everyone benefits when new treatments get to the pharmacy faster.

INDUSTRY NOTE: Drug Recalls

Recalls happen. The diabetes drug troglitazone, which was marketed as Rezulin in the US, caused acute liver failure in some patients. The FDA asked Parke-Davis, the manufacturer, to remove the drug from the market in 2000. The manufacturer agreed. Similarly, the cholesterol-lowering statin, Baycol caused at least 52 deaths from kidney failure and was withdrawn from the market in 2001.

Sometimes independent researchers analyze clinical trials after drug approval. **Meta-analysis** involves compiling data from several studies of the same drug. Some experts question the value of a purely statistical analysis since it doesn't take into account a trial's experimental design. Nonetheless, meta-analysis has profoundly impacted drug marketing and safety awareness.

GOING FURTHER: Meta-Analysis

Meta-analysis has called into question the safety of the diabetes medicine Avandia, implicating it in increased risk of heart attacks. Of more than 14,000 patients taking Avandia, 86 suffered a heart attack during the study. This compares to 72 patients out of more than 11,000 who took a "standard of care" drug like metformin or a placebo. In this case, the increased risk of heart attack seems small. Doctors and patients need to carefully weigh increased drug risk against likely benefit.

More Paperwork: Patents

In the United States, drug companies typically apply for a patent early in drug discovery. Patent protection expires twenty years from the date of filing, so its practical length varies depending on the duration of clinical trials. Expediting clinical testing is critical to drug companies. Still, some drugs win approval with just a few years of patent protection left.

Because of this, the FDA grants each new drug a period of exclusivity. This means that the holder of an approved new drug application receives protection from competition. For small molecule drugs, the exclusivity period is five years, plus a six-month extension for pediatric products. Orphan drugs—developed for a patient population of less than 200,000—receive seven years of exclusivity. Biologic drugs are granted 12. This longer timeframe acknowledges the substantially greater manufacturing costs associated with biologics.

BIOPHARMA INNOVATION: Rare Disease

Drug companies historically focus on "big" health problems such as cancer and heart disease. That's why in the early 1980s, the federal government passed the Orphan Drug Act to provide incentives to develop treatments for rare illnesses. These are defined as afflicting fewer than 200,000 people in the US. Such diseases include multiple myeloma, Huntingdon's disease, Lou Gehrig's disease, cystic fibrosis, Parkinson's disease, and phenylketonuria. These drugs obviously have fairly small potential markets. The legislation introduced tax breaks and an extra two years of market exclusivity, making orphan drug research more attractive. Since 1983, about 250 orphan drugs have made it to market. Prior to the law, fewer than 10 such products earned approval.

That Special Something

Certain drugs get special consideration throughout FDA approval. Such therapies are eligible for regulatory designations that speed up review and get a product to market more quickly. They must first meet two criteria. 1) A drug must target a serious condition that likely results in death or significantly impairs daily living. Examples include cancer and Alzheimer's disease. 2) The drug also needs to address a major, unmet medical need. Either no medicine exists or current therapy has safety issues.

The FDA's special designations include:

- **Accelerated Approval** allows drugs to go forward using surrogate instead of clinical endpoints. Surrogate endpoints, such as lowered blood pressure or reduced tumor size *predict* rather than demonstrate clinical benefits. In these cases, pharmaceutical companies run post-market studies to verify the anticipated effect.

- **Priority Review** means the FDA aims for a decision within six months.

- **Fast Track** is based on preclinical or clinical data that suggests the product addresses an unmet medical need. The designation enables developers to communicate more often with the FDA. The agency provides guidance on clinical trial design and process, which helps resolve questions or issues quickly. Designees also qualify for Accelerated Approval, Priority Review, and Rolling Review—in which developers submit each section of an NDA or BLA as they finish, rather than all at once.

- **Breakthrough Therapy** designates drugs that may greatly improve patient health. The bar is set high to join this privileged group, though. It requires preliminary clinical evidence of effectiveness. Once granted, breakthrough designees receive fast track advantages. The FDA also gives intensive regulatory guidance as early as Phase 1. Companies get an "organizational commitment involving senior FDA managers," according to the FDA website.

What's in a Name: Generic

The term **generic** refers to a small molecule drug that is an exact molecular copy of an innovator drug. If a generics manufacturer demonstrates molecular equivalence, then they may rely on the safety and efficacy data from the original drug manufacturer when filing an **Abbreviated New Drug Application (ANDA)**. The generic's **formulation** may differ slightly from that of the innovator compound. If it does, clinical data may be required to demonstrate the generic shares equivalent **bioavailability** with the original. Bioavailability is the portion of the drug that has an effect on the disease. If the ANDA is approved, the generic drug manufacturer receives a 180-day exclusivity period in which it is the sole company that can sell the generic. When the exclusivity period expires, other generics manufacturers can begin manufacturing their versions. Such robust competition keeps the price of generic drugs down.

Tricky Terms

A drug **formulation** refers to the different chemical substances, including the active drug, that go into the final medicinal product. Additives may include stabilizing agents, bulking agents, and solubilizing agents. Any can affect the active ingredient's **bioavailability,** the total amount that reaches the target tissue and the time it takes to arrive.

What's in a Name: Biosimilar

Large molecule drugs are structurally very complex in comparison to small molecule drugs. It's impossible to demonstrate that a large molecule drug produced by a company other than the innovator company, potentially using different cell lines and/or processes, is identical to the original therapeutic. Thus the term generic isn't used when referring to copies of biologic drugs. Instead researchers and drug companies call them **biosimilars**.

Biosimilars gained marketing approval in the EU in 2006. Three years later, Congress passed the Biologics Price Competition and Innovation Act, which gave the FDA the authority to approve biosimilar drugs. The law left the scientific guidelines regarding testing and approval up to the agency itself.

Similar to EMA parameters, the FDA guidance directs makers of biosimilars to demonstrate that they are highly similar to their **reference product**, the already-approved drug. The degree of similarity demonstrated in the analytical studies dictates the need for animal safety testing and human clinical trials. In most cases, some degree of evaluation turns out to be necessary. Any reduction in preclinical or clinical testing has to be scientifically justified by the sponsoring company. As with any new drug, companies are required to establish vigorous post-market safety monitoring plans.

In this chapter, we've outlined what it takes to get medicine from the lab to the market. The next describes large-scale manufacturing of biologic drugs, also known as biomanufacturing.

CHAPTER 9
Biomanufacturing

Biomanufacturing, as the name suggests, creates biological products from living cells. Biomanufacturing plants turn out a staggering array of products: enzymes, vaccines, therapeutic antibodies, antibiotics and more. Manufacturing consists of making very large quantities biological "stuff." For many products, it's the penultimate step in what can be a convoluted journey through the **product pipeline** to the market place.

More than most, manufacturers of biotech products must contend with regulatory issues. For example, Nike never worries about the FDA's **cGMP** (Current Good Manufacturing Practices) guidelines.

This chapter introduces regulatory concerns, but more so, overviews biotech manufacturing in general.

The Big Biomanufacturing Picture

It may be a little counterintuitive to think that anything good comes from bacteria. And yet, insulin! Enzymes! These and many other biomanufactured products make people's lives better every day.

How do companies get an enzyme from living bacterial cells? What follows is a basic model of biomanufacturing. The cell type used depends on the product. Proteins with a relatively simple structure, such as human insulin, can be made in bacterial cells. Structurally fancier proteins, such as monoclonal antibodies, need correspondingly more complex mammalian cells. Penicillin and other antibiotics, important as they are to human health, arise from lowly fungal cells.

Engineering bacterial cells to make therapeutic proteins begins by isolating the gene that encodes for that protein. Recombinant DNA technology puts the gene of interest in the appropriate plasmid which is a circular piece of bacterial DNA. This combination of a gene of interest and the bacterial plasmid yields a recombinant plasmid. (For more on recombinant technology, flip back to Chapter 5.) The newfangled plasmid gets transferred into cells, or, what could be called manufacturing cell lines. Cell lines that manufacture biologics can be bacterial cells, mammalian cells, insect cells, plant cells, and fungal cells.

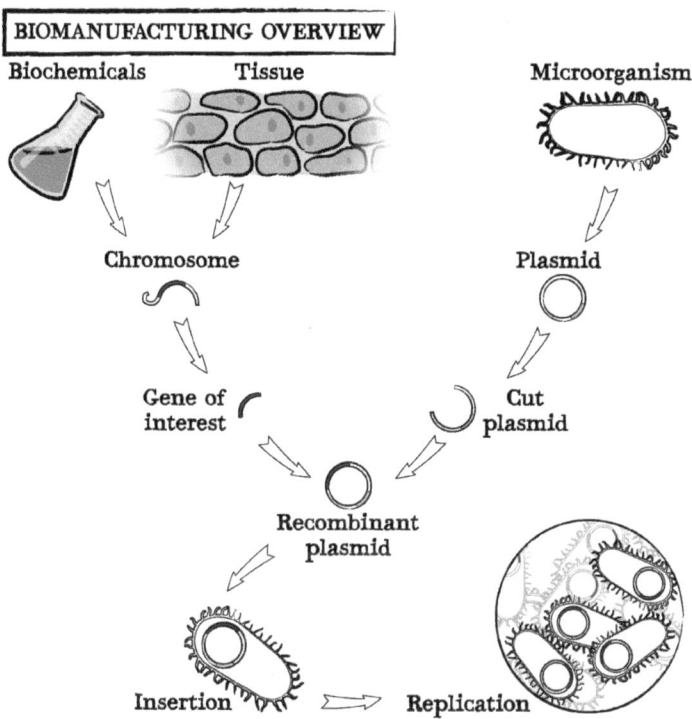

What comes next? The manufacturer needs more of the recombinant cells, so it's time to culture them. Once they start growing and dividing, the recombinant cells multiply and make recombinant protein from the new gene. This initial culture provides a small-scale batch of cells with the newly-inserted gene of interest.

Bigger Better Faster More!

Once the new and improved cells are growing nicely, the next stage is **scale-up**. Ever hear of a **benchtop bioreactor**? A **pilot scale bioreactor**? These doohickeys make biomanufacturing possible. They're essentially vats in which vast amounts of cells grow.

Like good gardeners, manufacturers must maintain proper growing conditions. No sun or rain here. Instead, the little protein factories need good nutrition, proper pH levels, sufficient oxygen, and adequate warmth. Bioreactors make scale-up possible. The bacteria are closely monitored and ambient conditions adjusted as needed. After successful production in the pilot scale reactor, the cells are transferred into industrial-scale bioreactors that use much bigger, sometimes humungous vessels. Once in a while these puppies hold a hundred thousand liters. Now that's scale-up!

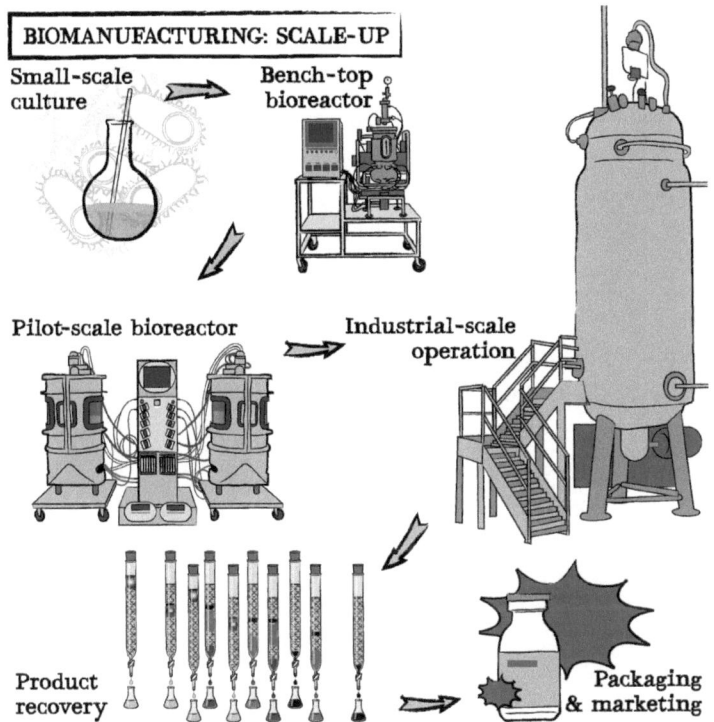

Once cells in a large-scale bioreactor reach maximum density, it's time to collect, or recover them. The cells sometimes secrete the desired protein into its growth media. Other times, the cells must be harvested and broken open—*lysed*—to release the protein goodies. Either way, the product needs to be purified before it's packaged as a therapeutic.

GOING FURTHER: Favorite Cells

We all play favorites, including research scientists. When it comes to mammalian cell culture, scientists love CHO (Chinese Hamster Ovary) and mouse myeloma Non-Secreting (NS0, pronounced "NS zero") cells. CHOs and NS0s have many advantages. They're relatively "easy" to engineer, have well-established and understood cell lines, grow nicely in volume, and yield lots of recombinant proteins. What's not to love? These handy cells process proteins much like human cells. The proteins are then likely to be folded likewise. They're also likely to have similar or identical **post-translational modifications**, such as the addition of sugar groups at specific locations called **glycosylation**, which a properly functioning drug needs in order to work. NS0s and CHOs have special FDA standing as **generally-regarded-as-safe (GRAS)** cells. Drug companies can use them to manufacture therapeutic proteins without first demonstrating their safety.

Full Speed Ahead

Going big in biomanufacturing can be difficult and time-consuming. Bacterial cells replicate quickly. That means scale-up sometimes takes less than a week. With mammalian cells, scale-up may take more than a month.

Campaign Season

The manufacturng process from start to finish is often called a "campaign." It's usually divided into two main parts: **upstream** and **downstream processing**. Upstream processing pertains to making the product. Downstream describes recovery, purification, formulation, and packaging.

BIOPHARMA INNOVATION: Formulation

During research and development, the company comes up with initial production methods on a small scale. Researchers eventually need to determine a final, viable physical formulation. Possibilities abound: tablet, inhaler, liquid, patch, cream? As mentioned previously, most biologics must be injected or administered intravenously; small molecule drugs typically take tablet form. Using data from earlier production steps, biomanufacturers establish the best way to produce the particular formulation of their product for its intended market.

Cell Banks

Upstream processing begins with the cells bioengineered to make the protein. Once they have the desired cell line, researchers **cryopreserve** it. They freeze many many cells in vials, creating a **cell bank**. Cell banking is two-tiered. There's a **master cell bank (MCB)** and a **working cell bank (WCB)**. The WCB originates from one vial of cells from the MCB. The WCB provides batches of product during scale-up process. Using the same stock of cell line reduces the risk of mutation. The MCB functions as a reserve, meant to last for the product's lifetime. This helps ensure the production of a consistent product, even decades after its launch. That's because the manufacturer continually goes back to the original source of cells: the MCB. Companies usually maintain MCB in three different geographic locations to protect against their loss.

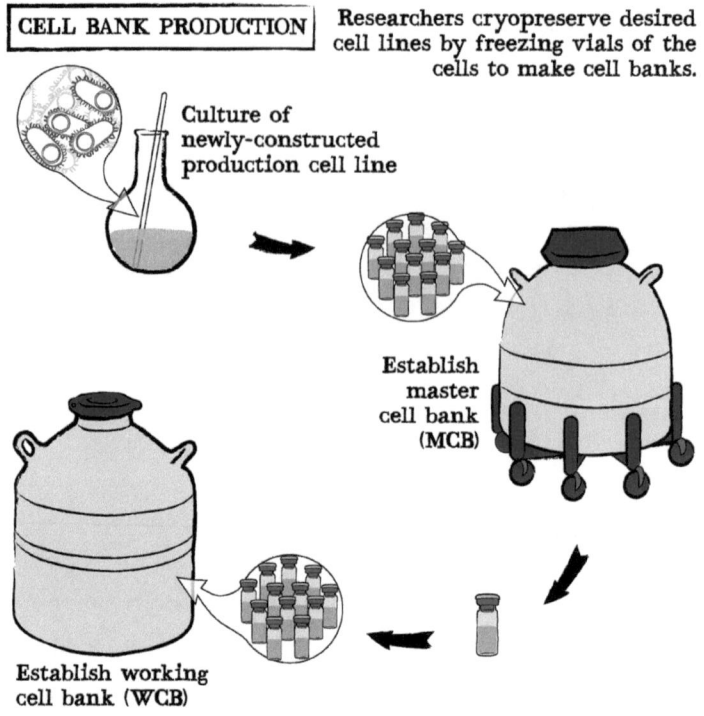

Scale-Up and Manufacturing Process

A production campaign begins with a withdrawal from the WCB. Thawed cells go into small benchtop bioreactors of "just the right stuff" including the right amount of oxygen, nutrients and pH level. By gradually transferring growing cells into successively larger growth vessels the manufacturer accomplishes, you guessed it, scale-up. Happy cells keep dividing and making more product. Toward the end of scale-up, cells are percolating in enormous vessels. For mammalian cells, bioreactors typically hold *fifteen* to *twenty thousand* liters. For a bacterial cell line, the vats contain up to a *hundred thousand liters* of a bacterial cell line. It bears repeating: now that's scale-up!

Scientists test cell viability, product concentration, and product activity every step of the way. They also monitor the physical surroundings. During early scale-up, technicians monitor the culture manually. The process becomes automated when the culture gets big enough for bioreactors.

It's critical during scale-up and manufacturing to check the cultures for contamination by bacteria, yeast, or other microorganisms. One bad apple (or grain of yeast) spoils the whole batch, costing millions of dollars and countless hours. Companies follow rigorous protocols to maintain **aseptic** conditions constantly.

Responsibility for monitoring falls to the **Quality Control (QC)** and **Quality Assurance (QA)** departments within the drug company. The QC department routinely tests the product to ensure that scale-up and manufacturing procedures meet standards. The QA department makes sure that quality objectives get met. It also handles the paperwork, reporting regularly as a product gets closer to marketing. These oversight practices necessarily include meeting FDA cGMP guidelines.

Harvest Time

In a product's downstream phase, it gets isolated from the cellular nursery. Extracting proteins from a cell or from *within* a cell requires special protocols. They usually involve bursting open the cells to release the therapeutic protein. Scientists then purify it, separating it from the other contents in the extract. Proteins excreted from the cell, such as monoclonal antibodies, are easier to isolate because breaking apart the cell and extracting the contents is unnecessary.

Column Chromatography

Purification often involves **column chromatography**, which separates proteins and other molecules from complex mixtures by characteristics such as size, shape, or electrical charge. The cell extract typically gets processed stepwise, using different column types.

Chapter 9: Biomanufacturing

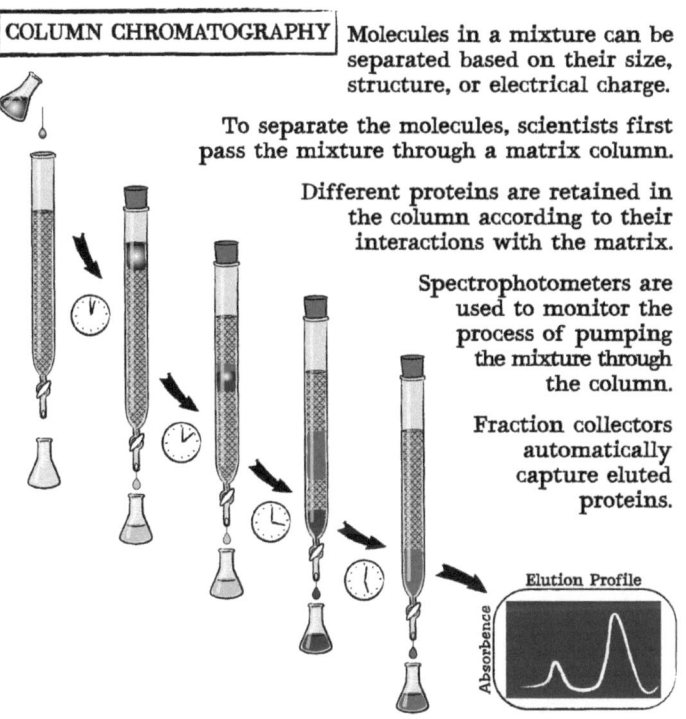

COLUMN CHROMATOGRAPHY — Molecules in a mixture can be separated based on their size, structure, or electrical charge.

To separate the molecules, scientists first pass the mixture through a matrix column.

Different proteins are retained in the column according to their interactions with the matrix.

Spectrophotometers are used to monitor the process of pumping the mixture through the column.

Fraction collectors automatically capture eluted proteins.

Elution Profile

Proteins are usually separated using **column chromatographs** that contains a **matrix**. The proteins are passed through the column under gravity and moderate pressure (fast protein liquid chromatography aka **FPLC**) or high pressure (high performance liquid chromatography aka **HPLC**) to increase speed and resolution. The FPLC method is generally the most effective and most often used.

Some proteins flow straight through the column; others remain inside the matrix or follow later (known as **elusion**). Elusion is scientist-speak for a particular kind of washing off. Scientists observe protein elution from the column using a **spectrophotometer**, which measures proteins' ability to absorb UV light. It provides an elution profile, which enables scientists to identify when proteins emerge from the column. **Fractions** corresponding to the different elution times are collected automatically in a fraction collector.

More Chromotography: Ion Exchange

Ion exchange chromatography separates proteins based on charge. Some amino acids, such as glutamate and aspartate, have a negative charge at neutral pH. Others, such as arginine and lysine, are positively-charged. Therefore, proteins have different charges depending on their composition. The ion exchange matrix contain beads coated with a negatively- or positively-charged chemical to separate the wheat from the chaff.

In a chromatography matrix of positively-charged beads (called **anion exchange**), positive and neutral proteins flow straight through the matrix. Negative proteins are retained, because they're attracted to the positively-charged coated beads. Scientists then elute the proteins from the matrix with a **gradient salt solution**. The solution causes **ions** to exchange with and displace proteins that **adsorbed** to the beads. Displaced proteins exit the column. By gradually increasing the solution's salt concentration, scientists make the protein molecules elute from the column according to charge strength. Molecules with a very strong ionic interaction require more salt and elute later in the gradient. Everything that washes out of the column is collected in a numbered series of test tubes as samples. These are fractions.

Still More: Affinity Chromatography

Affinity chromatography works on a similar principle to ion exchange chromatography. Here, the matrix separates protein molecules by shape, not charge. Only proteins with a "pocket" complementary to the unique shape attached to the bead bind to the column matrix. After the unbound proteins have been washed away, the bound proteins can be eluted with a soluble form of the shaped chemical that replaces the bound form.

And One More: Size Exclusion Chromatography

Size exclusion chromatography, commonly called gel filtration, works by a different principle. It separates proteins by size; larger proteins pass through the matrix faster than smaller ones. The matrix contains porous beads with pores of a defined size. Large proteins cannot get into the beads and flow around them. Smaller proteins *can* enter the beads and pass through, taking longer to reach the column's bottom than the big dudes. Different-sized proteins thus gradually get separated from one another and are collected once they've passed through the column. For gel filtration chromatography columns to work effectively, they have to be much longer and thinner than other column types to enable resolution of different size proteins.

Formulation, Fill...

After purification, the protein product can be formulated. The company may use product **excipients**, which are pharmacologically inactive ingredients designed to enhance the drug product. They may include color additives, time release factors, and bulking agents such as mannitol. Manufacturers may also add stability agents such as anti-oxidants, buffers, and surfactants (to decrease clumping).

...and Finish

Finally, the company establishes product fill concentration, and determines the labeling and packaging for large-scale distribution. With that, the biomanufacturing process is complete.

In this book, we have introduced some of the key scientific concepts underlying the biotech industry. Our next book, "The Biopharmaceutical Primer: A Guide to Understanding Advanced Therapies," explores how these concepts drive innovative new therapies.

CONCLUSION

In the preceding chapters, we have taken an in-depth look at the biotech industry, and in particular, the science that drives it. From cell structure to protein structure; gene expression to genetic variation and genetic engineering; the human immune response to the production of antibodies for biotech application; and finally drug discovery, drug development, and biomanufacturing—we have discussed the key concepts and technologies that impact current biotechnology developments.

It is our hope that you will use this book continuously as a reference to support your growth as a biotechnology professional. Although the industry itself is constantly changing, these fundamental concepts upon which it is built will remain important for years to come—and decision-makers who understand these fundamentals will be better able to evaluate and predict new trends.

More than anything else, we hope that your understanding of the science behind biotechnology will serve to increase your enthusiasm for this exciting and truly life-changing industry.

> Although the industry itself is constantly changing, these fundamental concepts upon which it is built will remain important for years to come...

GLOSSARY

A

Abbreviated New Drug Application (ANDA): An application for a US generic drug approval for an existing licensed medication or approved drug.

Action Potential: In physiology, an action potential is a short-lasting event in which the electrical membrane potential of a cell rapidly rises and falls, following a consistent trajectory. Action potentials occur in several types of animal cells, called excitable cells, which include neurons, muscle cells, and endocrine cells, as well as in some plant cells.

Active Site: The portion of an enzyme that attaches to the substrate.

Adaptive: Providing, contributing to, or marked by adaptation.

Adaptive Immune Response: The response of antigen-specific lymphocytes to antigen.

Adaptive Immune System: The response of antigen-specific lymphocytes to antigen, and includes the development of immunological memory.

Adenine: One of the four nucleotide bases that make up DNA.

Adsorbed: Attracted to and maintained on the surface of a solid surface, as in chromatography.

Affinity Chromatography: A chromatographic method that makes use of the specific binding of one molecule for another.

Allele: One of a number of different forms of a gene. Each person inherits two alleles for each gene, one allele from each parent. These alleles may be the same or may be different from one another.

Glossary

Amino Acid: One of twenty different molecules that combine to form proteins. The sequence of amino acids in a protein determines the protein's structure and function.

Angiogenesis: Growth of a network of blood vessels that penetrates into cancerous growths, supplying nutrients and oxygen and removing waste products.

Anion Exchange Chromatography: A type of column chromatography in which positively-charged proteins are captured by a negatively-charged chromatography matrix.

Antibody: A protein produced by the immune system that binds to a specific antigen.

Anticodon: A specific three-nucleotide sequence in transfer RNA that is complementary to a codon (a three-nucleotide sequence in messenger RNA) that specifies an amino acid in protein synthesis. When a codon and anticodon bind (because they are complementary strands) the amino acid attached to the transfer RNA is able to connect to the growing amino acid strand, which eventually forms a protein.

Antigen: A foreign substance which, when introduced into the body, stimulates an immune response.

Antigen Presenting Cell: A highly specialized cell that is able to process antigens and display their peptide fragments on the cell surface together with molecules required for lymphocyte activation.

Antiserum: The fluid component of clotted blood from an immune individual that contains antibodies against the molecule used for immunization.

Apoptosis: The process of cell self-destruction.

Aseptic: Free from contamination caused by harmful bacteria, viruses, or other microorganisms.

Assay: A test.

Atom: A particle, made up of a nucleus and one or more orbiting electrons, which is the basic unit of a chemical element.

ATP: Adenosine triphosphate; a compound used by cells to store energy and to fuel metabolic processes.

B

B-Cell: An antibody-producing cell of the immune system.

B-Cell Receptor: The cell-surface receptor of B cells for specific antigen.

Bacterium: A group of small, single-celled, prokaryotic microorganisms.

Base: One of the molecules - adenine, guanine, cytosine, thymine, or uracil - which form part of the structure of DNA and RNA molecules. The order of bases in a DNA molecule determines the structure of proteins encoded by that DNA. See nucleotide.

Base Pair (bp): Two complementary nucleotide bases joined together by chemical bonds. The base adenine pairs with thymine, and guanine pairs with cytosine.

Basophil: White blood cells that participate in the inflammatory response.

Benchtop Bioreactor: A small bioreactor, typically with volumes of three liters or less.

Bioactive: Having an effect on a biological system.

Bioavailability: The proportion of a drug or other substance which enters the circulation when introduced into the body and so is able to have an active effect.

Biofortification: The idea of breeding crops to increase their nutritional value. This can be done either through conventional selective breeding, or through genetic engineering. Biofortification differs from ordinary fortification because it focuses on making plant foods more nutritious as the plants are growing, rather than having nutrients added to the foods when they are being processed.

Biologics: Products of living organisms or cells used in the treatment or management of a disease.

Biologics Licensing Application (BLA): An application for marketing approval for a biologic drug; submitted to the FDA upon successful completion of Phase III clinical trials.

Biomanufacturing: The use of living cells to produce a biological product. An example is a therapeutic protein.

Biomarker: A physiological event or molecule that can be measured. Examples include the presence or absence a protein or a mutated gene. Biomarkers are often used to indicate the presence or progression of a disease.

Bioinformatics: Bioinformatics is an interdisciplinary field that develops methods and software tools for understanding biological data. As an interdisciplinary field of science, bioinformatics combines computer science, statistics, mathematics, and engineering to analyze and interpret biological data.

Biopharmaceutical: Drugs, either chemical compounds or biologics, made using the method of rational drug design.

Bioreactor: The usually stainless steel tank used to grow living cells that produce biologics. A bioreactor can range in size from a few liters up to 100,000 liters.

Glossary

Bioremediation: A technique used to remove or neutralize contaminants in a particular environment using living organisms.

Biosciences: A term meant to encompass both biotech and pharmaceutical companies.

Biosimilar: A biologic deemed highly similar to an already approved biologic medicine, known as a reference product. This similarity is confirmed through a number of tests, both in the lab and in clinical research.

Biotechnology: The use of cellular and biomolecular processes to solve problems and make useful products.

C

Carriers: A person who has inherited a genetic trait or mutation, but who does not display that trait or show symptoms of the disease.

Catalyzed: Caused or accelerated a reaction by acting as a catalyst.

CCR5 Gene: C-C chemokine receptor type 5 gene codes for a protein on the surface of white blood cells that is involved in the immune system as it acts as a receptor for chemokines. Many forms of HIV, the virus that causes AIDS, initially use CCR5 to enter and infect host cells.

Cell: The basic subunit of any living organism, typically containing at a minimum genetic material, an energy-producing system, and protein-making machinery, all surrounded by a membrane.

Cell Bank: A uniform population of cells, stored under defined conditions, typically frozen at -80 degrees Celsius or colder. The assumption is that each vial of cells is comparable, and may be used in a consistent manner after being thawed. See Master Cell Bank and Working Cell Bank.

Cell Line: A cell culture developed from a single cell and therefore consisting of cells with a uniform genetic makeup.

Cell Lysate: The cellular debris and fluid produced by brakeing open (lysing) a cell.

Cell Membrane: Barrier made out of fats and proteins that separates the inside of the cell from the outside.

Cell Wall: A stiff covering around the plasma membrane of certain non-animal cells, such as plants and many types of bacteria.

Channel Protein: A protein that spans the cell membrane, allowing substances to pass through form the outside to the inside of the cell.

Chemokine: Signaling molecules that are involved in the activation and migration of immune system cells. Chemokine signaling plays a key role in the inflammatory response.

Chimeric Antibody: Antibodies whose gene sequence consists of DNA from two different species. Typically, the term refers to antibodies whose DNA is between 10% and 25% mouse origin, with the remaining sequence being of human origin.

Chromosome: A long strand of DNA found within cells. Chromosomes contain both genes and regions of DNA that do not code for proteins.

Cleave: To split a molecule by breaking a particular chemical bond.

Clinical Endpoint: In a clinical research trial, a clinical endpoint generally refers to occurrence of a disease, symptom, sign or laboratory abnormality that constitutes one of the target outcomes of the trial.

Clonal Selection: A process proposed to explain how a single B or T cell that recognizes an antigen that enters the body is selected from the pre-existing cell pool of differing antigen specificities and then reproduced to generate a clonal cell population that eliminates the antigen.

Codon: A set of three nucleotide bases in a DNA or RNA sequence, which together code for a unique amino acid. For example, AUG (adenine, uracil, guanine) codes for the amino acid methionine.

Column Chromatography: A laboratory technique for the separation of a mixture of proteins based on each protein's unique chemical and physical characteristics.

Combination: A product comprised of two or more FDA-regulated categories.

Combinatorial Chemistry: A technique for rapidly and systematically assembling different combinations of molecules to create tens of thousands of diverse compounds. Used in drug discovery screening assays to identify potential useful therapeutic candidates.

Combinatorial Libraries: Collections of chemical compounds, small molecules or macromolecules such as proteins, synthesized by combinatorial chemistry, in which multiple different combinations of related chemical species are reacted together in similar chemical reactions.

Companion Diagnostic: A diagnostic used by a physician to inform his prescribing decision.

Contract Research Organization (CRO): A company that conducts preclinical or clinical trials for another company on a contract basis.

Glossary

Control Group: The group in an experiment or study that does not receive the experimental treatment and is then used as a benchmark to measure how the experimental group does.

Copy Number Variation: A form of structural variation that manifest as deletions or duplications in the genome. For example, the chromosome that normally has sections in order as A-B-C-D might instead have sections A-B-C-C-D (a duplication of "C") or A-B-D (a deletion of "C"). Cells with CNVs have abnormal or, for certain genes, normal variations in their copy number.

Current Good Manufacturing Practices (cGMP): FDA guidelines governing the manufacturing of biopharmaceuticals.

Cryopreserve: The preservation of cells or whole tissues by cooling to low sub-zero temperatures.

Cytochrome P450: A liver enzyme which plays an important role in metabolizing drugs.

Cytokine: Proteins made by cells that affect the behavior of other cells.

Cytoskeleton: Network of protein filaments just underneath the cell membrane that gives the cell shape.

Cytosine: One of the four nucleotide bases that make up DNA.

Cytotoxic T-Cells: T-cells that can kill other cells, typically virus-infected cells or tumor cells.

Cytotoxins: Proteins made by cytotoxic T cells that participate in the destruction of other cells.

D

Deletion Mutation: One or more nucleotides is removed from a DNA sequence during the replication process.

Denature: Dramatic change in the conformation of a protein, usually unfolding caused by heat or exposure to chemical such as acids. Most often will result in loss of biological function.

De Novo: From the beginning.

Deoxyribonucleic Acid (DNA): The molecule that encodes genetic information. DNA is a double-stranded helix held together by bonds between pairs of nucleotides.

Deoxyribose: A type of sugar which is a component of DNA (Deoxyribonucleic Acid).

Diagnostic (Dx): A test used to identify a disease or disorder, or to monitor the progression of treatment. Routine diagnostics are broad screening tools, whereas a specialty diagnostic screens for a specific disease.

Directed Evolution: A method used in protein engineering that mimics the process of natural selection to steer proteins or nucleic acids toward a user-defined goal.

DNA Sequence: The precise order of nucleotides within a DNA molecule. It includes any method or technology that is used to determine the order of the four bases—adenine, guanine, cytosine, and thymine—in a strand of DNA.

Dominant: A disease-associated gene is dominant if only one copy of it needs to be present to cause disease.

Double Helix: The shape of the DNA molecule.

Downstream Processing: The phase of a biomanufacturing campaign that consists of harvesting, purifying, and formulating the product.

DNA Helicase: The enzyme that separates the two strands of a DNA molecule prior to replication.

DNA Ligase: An enzyme that acts as a molecular glue, gluing two pieces of DNA together.

DNA Polymerase: The enzyme that replicates DNA.

DNA Repair Genes: A collection of genes that code for proteins that enable a cell to identify and correct damage to the DNA molecules that encode its genome.

Domain: Portion of a protein that has its own structure.

Drug Candidate: A small molecule or biologic that is being tested for its therapeutic potential.

Drug Development: The process of testing therapeutic molecules for safety and efficacy in animals and humans, and developing appropriate formulation, delivery, and manufacturing methods.

Drug Discovery: The process of identifying molecules with a therapeutic effect against a target disease.

Drug Target: Organ, tissue, or molecule involved in a disease that is modified or affected by a potential therapeutic.

E

EC50: Drug concentration at which one-half of the maximum desired effect is observed.

E. coli: Common bacterium that has been studied intensively by geneticists because of its small genome size, normal lack of pathogenicity, and ease of growth in the laboratory.

Efficacy: The ability of a substance to produce a desired clinical effect.

Glossary

EGFR: Epidermal growth factor receptor. Transmits proliferation signals to many different cell types. Mutations in this protein are associated with different types of cancer.

Electroporation: The act of applying an external electrical field to a cell membrane in order to increase its permeability. The technique is used as a way of introducing DNA into bacterial cells.

Electrophoresis: A method of separating large molecules (such as DNA fragments or proteins) from a mixture of similar molecules. An electric current is passed through a medium containing the mixture, and each kind of molecule travels through the medium at a different rate, depending on its electrical charge and size.

Elute: To wash off or remove adsorbed material from a solid support.

Endogenous: Having an internal cause or origin.

Enzyme: A protein that enables a biochemical reaction in a cell.

Eosinophil: White blood cells thought to be important chiefly in the defense against parasitic infections.

Epidermal Growth Factor (EGF): A protein which binds to the epidermal growth factor receptor and stimulates cell division.

Epigenetic: Relating to or arising from nongenetic influences on gene expression.

Epitope: The structural feature of an antigen molecule to which an antibody binds.

Erythrocyte: Red blood cells.

Eukaryote: Cell or organism with membrane-bound, structurally discrete nucleus and other well-developed subcellular compartments.

European Medicines Agency: European Union agency for the evaluation and supervision of medicinal products.

Evolution: The natural selection of beneficial changes.

Ex vivo: Pertaining to a biological process or reaction taking place outside of a living organism.

Excipient: Anything present in a drug product that is not the active ingredient. Includes components such as bulking and stabilizing agents, preservatives, salts, solvents, and water.

Expression Cloning: A technique in DNA cloning that uses expression vectors to generate a library of clones, with each clone expressing one protein. This expression library is then screened for the property of interest and clones of interest recovered for further analysis. An example would be using an expression library to isolate genes that could confer antibiotic resistance.

F

Formulation: The process in which different chemical substances, including the active drug, are combined to produce a final medicinal product.

FPLC: Fast Protein Liquid Chromatography

Fractions: A separate portion of a mixture, often used to describe the part that contains a particular molecular species.

Fragment Crystallizable (FC) Region: The tail region of an antibody that interacts with cell surface receptors called Fc receptors. This property allows antibodies to activate the immune system.

G

G Protein-Coupled Receptor (GPCR): A type of receptor protein that is targeted in over 25% of all biotech drugs.

Gamete: Reproductive cells; the ovum or sperm.

Gene: A length of DNA that codes for a particular protein.

Gene Amplification: The multiple replication of a section of the genome, which occurs during a single cell cycle and results in the production of many copies of a specific sequence of the DNA molecule.

Gene Expression: The process by which the information in a gene is used to create proteins.

Gene Of Interest: A gene that is under study by a researcher as a potential recipe for a biologic drug or because of its relevance to disease.

Gene Product: The protein produced by a gene.

Generally Regarded As Safe (GRAS): A special status afforded by the FDA to ingredients and methods that have a proven, longstanding history of causing no harm to humans or animals.

Generic Drug: A small molecule drug that is an exact molecular copy of an innovator drug.

Genetic Code: Set of rules by which the information encoded in DNA is translated into proteins.

Genetic Engineering: Altering the genetic material of cells or organisms in order to make them capable of producing new substances or performing new functions.

Genetic Variation: Differences in DNA sequence that occurs between individuals.

Genetically Engineered Organism: An organism whose DNA has been altered using genetic engineering techniques.

Genetically Enhanced Organism: An organism whose DNA has been altered using genetic engineering techniques.

Glossary

Genetically Modified Organism (GMO): An organism whose DNA has been altered using genetic engineering techniques.

Genome: All of the genetic material in the chromosomes of a particular organism.

Genomics: The branch of molecular biology concerned with the structure, function, evolution, and mapping of genomes.

Genotype: The genetic constitution of an individual organism.

Germ Cell: A cell containing half the number of chromosomes of a somatic cell and able to unite with one from the opposite sex to form a new individual; a gamete.

Glycosylation: Adding one or more carbohydrate molecules onto a protein after it has been built by the ribosome.

Golgi Body: Cellular organelle which sorts and sends proteins to their appropriate location within the cell.

Gradient Salt Solution: A gradual change in the concentration of solutes in a solution as a function of distance through a solution

Granulocyte: A type of white blood cells characterized by the presence of small particles in their cytoplasm.

Growth Factor: A substance, such as a vitamin or hormone, which is required for the stimulation of growth in living cells.

Growth Factor Signaling: The process of transmitting a chemical signal from a cell surface receptor to the nucleus of a cell.

Growth Medium: A liquid or gell designed to support the growth of cells or microorganisms. There are different types of media for growing different types of cells.

Guanine: One of the four nucleotide bases that makes up DNA.

H

Half-Life: The amount of time it takes for 50% of a drug given to a patient to be eliminated or destroyed by natural processes.

Haplotype: Set of single nucleotide polymorphisms (SNPs) on a chromosome which are statistically associated.

Hematopoiesis: The generation of the cellular components of blood, including red blood cells, white blood cells, and platelets.

Hematopoietic Stem Cell: Stem cells found in the bone marrow that have the potential to develop into any of the different types of blood cells found in the body.

Hemostasis: Clotting of blood.

High Throughput Screening (HTS): The automated trial-and-error testing, typically using robotics, of very large sets of chemicals or materials.

Histone: Proteins that chromosomal DNA is wrapped around.

HPLC: High Pressure Liquid Chromatography

Human Anti-Mouse Antibody (HAMA): Antibodies generated by the human immune system when a human is injected with an antibody produced by a mouse.

Humanized Antibody: An antibody produced in a non-human species whose DNA sequence has been altered to make it more closely resemble a human antibody.

Hybridoma: Hybrid cell lines formed by fusing a specific antibody-producing B cell with a myeloma cell line.

I

Immune System Checkpoints: Regulators of the immune system. These pathways are crucial for self-tolerance, which prevents the immune system from attacking cells indiscriminately. Inhibitory checkpoint molecules are targets for cancer immunotherapy due to their potential for use in multiple types of cancers.

Immunological Memory: The ability of the immune system to quickly and specifically recognize an antigen that the body has previously encountered and initiate a corresponding immune response.

Inflammation: A non-specific immune defense by the body in response to injury or the presence of foreign particles.

Innate Immune Response: The early phase of the host response to infection in which a variety of mechanisms recognize and respond to a pathogen. Innate immunity is present in all individuals at all times, does not increase with exposure to a given pathogen, and does not discriminate between pathogens. Inflammation is an example of an innate immune response.

Innate Immune System: The "first line of defense" in immunity. Responds non-specifically to any invading pathogen.

Interferon: Cytokines that can induce cells to resist viral infection.

Interleukin: A generic term for cytokines produced by white blood cells. Specific interleukins may have either an inhibitory or a stimulatory effect on the immune system.

Intracellular Vesicles: Sometimes cells need to transport small amounts of material, like specific molecules. The process of surrounding small quantities of material, either from outside the cell or from some organelle is called pinocytosis. The small sacs that carry such material are called vesicles.

Ions: A positively or negatively charged atom.

Ion Channel: A channel protein through which ions are allowed to pass.

Glossary

Ion Exchange Chromotography: A type of chromatography that depends on adsorbing a charged protein to a matrix carrying the opposite charge.

In Silico: Studies done on a computer; for example, modeling the structure and function of a protein.

In Vitro: Pertaining to a biochemical process or reaction taking place in a test-tube as opposed to taking place in an organism.

In Vivo: Pertaining to a biological process or reaction taking place in a living organism.

Insertion Mutation: One or more nucleotides is added to the DNA sequence.

Investigational New Drug (IND): A drug which has gained FDA approval to be shipped across state lines, typically for clinical trials, but has not yet gained approval for marketing.

K

Kinase: An enzyme that transfers a phosphate group from ATP to a protein. Often, this results in activation of the recipient protein.

L

Large Molecule Drug: Another name for protein therapeutics. Large molecule drugs are too large to enter cells.

Leukocyte: General term for white blood cell.

Life Sciences: A term meant to encompass both biotech and pharmaceutical companies.

Ligand: Any molecule that binds to a specific site on a protein or other molecule.

Ligand-Gated Ion Channel Signaling: Ligand-gated Ion Channels (LGICs) are a group of transmembrane ion channels that open when a signal molecule (ligand) binds to an extracellular receptor region of the channel protein.

Ligation: The act of "gluing" two pieces of DNA, using the enzyme DNA ligase.

Liposome: Artificial lipid bilayer vesicle.

Lymphocyte: A white blood cell that is important in the body's immune response.

Lymph System: The system of lymphoid channels that drains extracellular fluid from the periphery.

M

Macrophage: A type of white blood cells that destroys pathogens by engulfing them.

Master Cell Bank (MCB): A culture of fully characterized cells distributed into separate vials, processed together in such a manner as to ensure uniformity. The master cell bank is usually stored at -80 degrees or colder (liquid nitrogen), and at two geographically distinct locations.

Matrix: In column chromatography, the solid phase to which proteins are adsorbed.

Maximum Tolerated Dose (MTD): Highest dose of a pharmacological treatment that will produce the desired effect without unacceptable toxicity.

Mechanism-Based Drug Design: The development of new therapeutics based on an understanding of the underlying disease mechanism.

Medical Device: An instrument, apparatus, implant, in vitro reagent, or other similar or related article, which is intended for use in the diagnosis of disease or other conditions, or in the cure, mitigation, treatment, or prevention of disease, or intended to affect the structure or any function of the body and which does not achieve any of its primary intended purposes through chemical action within or on the body.

Memory B-Cell: A B-cell that remembers the same pathogen for faster antibody production in future infections.

Messenger RNA (mRNA): The DNA of a gene is copied into mRNA molecules, which then serve as a template for the synthesis of proteins.

Meta-Analysis: The statistical procedure for combining data from multiple studies. When the treatment effect (or effect size) is consistent from one study to the next, meta-analysis can be used to identify this common effect.

Microbiome: The totality of microbes, their genomes, and environmental interactions in a particular environment.

Microinjection: The use of a glass micropipette to inject a liquid substance at a microscopic or borderline macroscopic level. The target is often a living cell but may also include intercellular space.

Microtiter Plate: A flat plate with lots of "wells" used as small test tubes.

Mitochondrion: An organelle within a cell that generates most of the cell's energy.

Molecular Cloning: A set of experimental methods in molecular biology that are used to assemble recombinant DNA molecules and to direct their replication within host organisms. The use of the word cloning refers to the fact that the method involves the replication of one molecule to produce a population of cells with identical DNA molecules.

Glossary

Molecule: Two or more atoms connected by chemical bonds.

Monoclonal Antibody (mAB): An antibody produced by a single clone of cells, which therefore consistently binds to the same epitope of an antigen.

Monocyte: Precursor to macrophages.

Monogenic Disease: A disease that can be linked to a mutation in one specific gene.

Multicellular: Having or consisting of many cells.

Multiple Ascending Doses: A group of patients receives multiple low doses of the drug, while samples (of blood, and other fluids) are collected at various time points and analyzed to acquire information on how the drug is processed within the body. The dose is subsequently escalated for further groups.

Mutation: A change, deletion, or rearrangement in the DNA sequence that may lead to the synthesis of an altered or inactive protein or the loss of the ability to produce the protein.

Myeloma: A cancer of plasma cells.

N

Natural Selection: A mechanism of evolution whereby members of a population with the most successful adaptations to their environment are most likely to survive and reproduce.

Neurotransmitters: Endogenous chemicals that enable neurotransmission. It is a type of chemical messenger which transmits signals across a chemical synapse, such as a neuromuscular junction, from one neuron (nerve cell) to another "target" neuron, muscle cell, or gland cell.

New Drug Application (NDA): An application for marketing approval for a small molecule drug; submitted to the FDA upon successful completion of Phase III clinical trials.

Nocebo Effect: A nocebo effect is said to occur when negative expectations of the patient regarding a treatment cause the treatment to have a more negative effect than it otherwise would have.

Non-Specific Immune Response: A non-specific immune cell is an immune cell (such as a macrophage, neutrophil, or dendritic cell) that responds to any foreign threat. Non-specific immune cells function in the first line of defense against infection or injury.

Nuclear Magnetic Resonance (NMR): A technique used to determine protein structure by measuring the absorption of electromagnetic radiation at a specific frequency.

Nuclear Membrane: The membrane surrounding the nucleus.

Nucleic Acid: One of the family of molecules which includes the DNA and RNA molecules.

Nucleotide: The "building block" of nucleic acids, such as DNA and RNA molecules. A nucleotide consists of one of five bases - adenine, guanine, cytosine, thymine, or uracil - attached to a sugar-phosphate group.

Nucleus: The membrane bound structure containing a cell's DNA found within all eukaryotic cells.

Nerve Growth Factor: A growth factor that binds to receptors on the surface of nerve cells, stimulating their growth.

Neutrophil: A major class of white blood cells. Play an important role in engulfing and killing foreign invaders.

O

Oncogene: A gene which in certain circumstances can transform a cell into a tumor cell.

Organelle: Membrane-bound structures in a cell that have specialized functions, such as mitochondria and the nucleus.

Orphan Drug: A drug developed for a condition that affects fewer than 200,000 individuals in the US.

Osteoblast: A cell responsible for bone formation.

Outcome-Based Reimbursement: A payment model in which insurance companies or other payors only pay for a drug or other type of treatment if it results in certain defined outcomes.

P

p53: A tumor suppressor gene. Mutations in p53 are linked to many different types of cancer.

Pathogen: A bacterium, virus, or other microorganism that can cause disease.

Peptide Bond: A type of chemical bond connecting amino acids.

pH: A measure of the acidity (hydrogen ion concentration) of a solution.

Phagocyte: A type of cell within the body capable of engulfing and absorbing bacteria and other small cells and particles.

Pharmacodynamics (PD): The study of the effect of a drug on the body; in particular, the effect of the drug as it relates to increasing dose.

Pharmacogenomics: The study of the role of genetics in drug response. It deals with the influence of acquired and inherited genetic variation on drug response in patients by correlating gene expression or single-nucleotide polymorphisms with drug absorption, distribution, metabolism and elimination.

Glossary

Pharmacokinetics (PK): The study of drug absorption, drug distribution within the body, drug metabolism, and drug excretion.

Pharmacovigilance: The pharmacological science relating to the collection, detection, assessment, monitoring, and prevention of adverse effects with pharmaceutical products.

Phase I Clinical Trial: Initial studies to determine the metabolism and pharmacologic actions of drugs in humans, the side effects associated with increasing doses, and to gain early evidence of effectiveness; may include healthy participants and/or patients.

Phase II Clinical Trial: Controlled clinical studies conducted to evaluate the effectiveness of the drug for a particular indication or indications in patients with the disease or condition under study and to determine the common short-term side effects and risks.

Phase III Clinical Trial: Expanded trials after preliminary evidence suggesting effectiveness of the drug has been obtained, and are intended to gather additional information to evaluate the overall benefit-risk relationship of the drug and provide and adequate basis for physician labeling.

Phase IV Clinical Trials: Clinical trials conducted to identify and evaluate the long-term effects of new drugs and treatments over a lengthy period for a greater number of patients. Phase IV research takes place after the FDA approves the marketing of a new drug. Through Phase IV clinical studies, new drugs can be tested continuously to uncover more information about efficacy, safety and side effects after being approved for marketing.

Phosphate: A chemical group consisting of an atom of phosphorous chemically bonded to four oxygen atoms.

Phosphorylate: To introduce a phosphate group into a molecule or compound.

Pilot Scale Bioreactor: An intermediate-scale bioreactor, typical volumes up to a few hundred liters.

Placebo Effect: A beneficial effect produced by a placebo drug or treatment, which cannot be attributed to the properties of the placebo itself, and must therefore be due to the patient's belief in that treatment.

Placebo Group: A control group that receives a sham treatment which is specifically designed to have no real effect.

Plasma Cell: An antibody-producing B-cell.

Glossary

Plasmid: A small, circular piece of DNA that is separate from the cell's genome. Plasmids are manipulated in the laboratory to deliver specific genetic sequences into a cell.

Platelets: Blood cells that are required for blood clotting.

Point Mutation: A single nucleotide change in a DNA sequence.

Polyclonal Antibody: A mixture of antibodies that recognize different epitopes on the same antigen; each antibody is produced by a different B cell.

Polymorphism: The condition of occurring in several different forms.

Polygenic Disease: A disease that results from interactions among two or more genes.

Polypeptide Chain: A chain of amino acids linked together through chemical bonds.

Post-Translational Modification: Changes made to a protein after it has been made by a cell.

Precipitated: The creation of a solid from a solution. When the reaction occurs in a liquid solution, the solid formed is called the 'precipitate'.

Precision Medicine: A medical model that proposes the customization of healthcare, with medical decisions, practices, and/or products being tailored to the individual patient. In this model, diagnostic testing is often employed for selecting appropriate and optimal therapies based on the context of a patient's genetic content or other molecular or cellular analysis.

Preclinical Studies: The testing of experimental drugs in the test tube or in animals - the testing that occurs before trials in humans may be carried out.

Primary Response: The adaptive immune response after initial exposure to an antigen.

Product Pipeline: The series of products developed and sold by a company, ideally in different stages of their life cycle.

Progenitor Cells: The more differentiated offspring of stem cells that give rise to distinct subsets of mature blood cells.

Prokaryote: Cell or organism lacking a membrane-bound, structurally discrete nucleus and other subcellular compartments. Bacteria are prokaryotes.

Promoter: A segment of DNA located in front of a gene, which provides a site where an enzyme can bind to the DNA molecule, to initiate transcription.

Glossary

Prostaglandins: Signaling molecules that have a variety of physiological effects, including smooth muscle contraction, platelet aggregation, and control of cell growth.

Protease: An enzyme that degrades proteins.

Protein: A biological molecule that consists of many amino acids linked together by peptide bonds. As the chain of amino acids is being synthesized, it is also folded into higher order structures. Proteins are required for the structure, function, and regulation of cells, tissues, and organs in the body.

Proto-Oncogene: Any gene capable of becoming a cancer-producing gene (an oncogene). Proto-oncogenes have important functions in the normal cell, but, by mutation or by the acquisition of genetic control elements from oncoviruses they can lose their normal regulatory functions and lead to uncontrolled multiplication.

PTEN: A human tumor suppressor proteins. As such, it regulates cell growth and division.

Q

Quality Assurance (QA): The quality systems and processes used to control every step of pharmaceutical manufacturing to ensure that the product meets all of its specifications and quality attributes, and that all steps were done and documented in compliance with cGMP.

Quality Control (QC): The system of testing that confirms and measures the quality of raw materials, process intermediates, final product and environmental samples.

R

R Group: In amino acids, the chemical group that varies and gives each amino acid its unique chemical and physical properties.

Randomized Double Blind Studies: Patients are randomly assigned to either the treatment or control group, and neither the patient nor the trial administers know which group is receiving the treatment.

Rational Drug Discovery: The development of new therapeutics based on an understanding of the underlying disease mechanism.

Readout: A visual record or display of the output from a computer or scientific instrument.

Real-World Evidence: Evidence obtained from real-world data (RWD), which are observational data obtained outside the context of randomized controlled trials (RCTs) and generated during routine clinical practice.

Receptor: A protein usually found on the surface of a cell that binds to a specific chemical messenger, such as a neurotransmitter or hormone.

Recessive: A disease-associated gene is recessive if two abnormal copies need to be present to cause disease.

Recombinant DNA (rDNA): DNA molecules that have been created by combining DNA from more than one source.

Reference Product: The single biological product, already approved by FDA, against which a proposed biosimilar product is compared.

Regulatory T-Cells: A subpopulation of T cells that modulate the immune system, maintain tolerance to self-antigens, and prevent autoimmune disease.

Research Support Company: A company who does not actively engage in drug discovery, but supplies all of the tools and technologies required.

Restriction Enzyme (RE): Proteins that recognize and cut DNA strands like molecular scissors along specific gene sequences.

Resolution: The dissipation of an inflammatory response.

Ribonucleic Acid (RNA): A nucleic acid similar to DNA but based on the sugar ribose and containing the nucleotides guanine, adenine, uracil, and cytosine instead of guanine, adenine, thymine, and cytosine, and typically single-stranded.

Ribosome: The cell structures within which protein synthesis occurs.

RNA Inhibition (RNAi): A biological process in which RNA molecules inhibit gene expression or translation, by neutralizing targeted mRNA molecules. Also referred to as RNA interference.

RNA Interference (RNAi): A technique used to block the expression of a particular protein. Also referred to as RNA inhibition.

S

Scale-Up: The process of slowly increasing the volume of a cell culture from a few milliliters to several thousand liters.

Secondary Response: The antibody response induced by a second exposure to the antigen.

Sequence: The order of nucleotides in a DNA or RNA molecule, or the order of amino acids in a protein.

Signal Transduction: The general process by which cells receive information from their environment.

Single Ascending Doses: Subjects are dosed in small groups called cohorts. Each member of a cohort might receive a single dose of the study drug or a placebo. A very low dose is used for the first cohort. The dose is then escalated in the next cohort if safety and tolerability allow. Dose escalation is stopped when maximum tolerability and/or maximum exposure is reached.

Glossary

Single Nucleotide Polymorphism (SNP): A difference in one base pair between two DNA sequences.

Small Interfering RNA: Small interfering RNA, sometimes known as short interfering RNA or silencing RNA, is a class of double-stranded RNA molecules, 20-25 base pairs in length, which operate within the RNA inhibition pathway.

Small Molecule Drug: A drug that is chemically synthesized in the lab. Small molecule drugs are small enough to enter cells.

Somatic Cell: A cell that is not an egg or a sperm cell.

Specific Immune Response: The immune response triggered by a specific pathogen.

Spectrophotometer: A machine used to measure the transmittance of ultraviolet light through a solution in order to determine the concentration of protein in that solution.

Stationary Phase: In column chromatography, the column matrix.

Statins: A class of drugs often prescribed by doctors to help lower cholesterol levels in the blood. By lowering the levels, they help prevent heart attacks and stroke. Studies show that, in certain people, statins reduce the risk of heart attack, stroke, and even death from heart disease by about 25% to 35%.

Start Codon: The three-nucleotide sequence that signifies the ribosome to start translating an mRNA sequence into a protein.

Stop Codon: The three-nucleotide sequence that signifies the ribosome to stop translating an mRNA sequence into a protein.

Stromal Cells: Connective tissue in the bone marrow and other tissue types.

Substitution Mutation: A single nucleotide is exchanged for a different nucleotide in a DNA sequence.

Substrate: A molecule on which enzymes act.

Surrogate Endpoint: In clinical trials, a surrogate endpoint (or surrogate marker) is a measure of effect of a specific treatment that may correlate with a real clinical endpoint but does not necessarily have a guaranteed relationship.

T

T-Cell: An immune system cell that recognizes specific pathogens based on the shape of its cell-surface receptor.

T-Cell Dependent Activation: Certain antigens can only activate B-cells with the help of T-cells.

T-Cell Receptor: Protein on the surface of T cells to which antigen binds, activating T cell.

T-Helper Cell: T cells that help to fully activate antibody-secreting plasma cells by secreting activating cytokines.

Tag SNP: A specific SNP that is used to identify a particular haplotype.

Target Validation: Determining if targeting a particular molecule thought to be involved in a disease mechanism will be a safe and effective means of therapy.

Template: The strand of DNA that is being used by the DNA polymerase to construct a new DNA molecule.

Tetramer: A protein composed of four subunits.

Therapeutic Window: Concentration range within which a drug is both effective and safe.

Thrombocytes: Blood cells that are required for blood clotting. Also called *platelets*.

Thymine: A compound which is one of the four constituent bases of nucleic acids. It is paired with adenine in double-stranded DNA.

Thymus: A small, irregular-shaped gland in the top part of the chest, just under the breastbone and between the lungs. *T-cells* complete their development in the thymus.

Tissue Culture: The growth in an artificial medium of cells derived from living tissue.

Transcription: The process during which the information in a length of DNA is used to construct an mRNA molecule.

Transduce: To convert a chemical signal from one form to another.

Transfection: The process of introducing DNA into eukaryotic cells.

Transfer RNA (tRNA): RNA molecules that bind to amino acids and transfer them to ribosomes, where protein synthesis is completed.

Transformation: A process by which the genetic material carried by an individual cell is altered by incorporation of external DNA into its genome.

Transformed: Bacterial cells that have had foreign DNA introduced.

Transgenic Organism: An organism whose genome has been altered by the incorporation of foreign DNA.

Translation: The process during which the information in mRNA molecules is used to construct proteins.

Tregitope: An epitope recognized by regulatory T-cells.

Trypsin: A protease found in the digestive system of many vertebrates, where it digests proteins.

Glossary

Tumor Suppressor Gene: A gene that protects a cell from one step on the path to cancer. When this gene mutates to cause a loss or reduction in its function, the cell can progress to cancer, usually in combination with other genetic changes.

U – Z

Unicellular: Organisms made up of only one cell.

Uracil: One of the four nucleotide building blocks of DNA.

Upstream Processing: The phase of biomanufacturing that consists of establishing cell banks and seeding and scaling up cell cultures.

Vasodilation: Dilation of a blood vessel.

Vector: A vehicle for the transfer of DNA from one organism to another. A plasmid is a common type of vector.

Working Cell Bank (WCB): A cell bank that is established from one of the master cell bank vials.

Xenotransplant: Transplantation of tissue or organs between organisms of different species, genus or family. A common example is the use of pig heart valves in humans.

X-Ray: A form of electromagnetic radiation.

X-Ray Crystallography: Technique for determining the three-dimensional arrangement of atoms in a molecule based on the diffraction pattern of X-rays passing through a crystal of the molecule.

Xenotransplantation: The therapeutic use of animal organs in people.

Printed in Poland
by Amazon Fulfillment
Poland Sp. z o.o., Wrocław